Finance Transformation

Finance Transformation: Leadership on Digital Transformation and Disruptive Innovation is a general and wide-ranging survey of finance transformation and emerging technologies. Finance and IT have long been important areas of any business, but recent technological developments are innovating and disrupting both. This book lays a path towards the benefits and away from potential risks. It covers the widest array of topics, from quantum computing to blockchain technology, from organisational culture and diversity to hybrid working, and from regulation to cybersecurity. Written by two vastly experienced industry professionals, this book includes real-life examples and up-to-date references. It will be of particular interest to business stakeholders, executives, and policymakers.

Aikta Varma is a finance leader with more than 16 years of progressive leadership experience at financial service (FS) firms, including Paragon Banking Group (FTSE 250 listed) and KPMG. She is a Fellow of the Institute of Chartered Accountants in England and Wales (ICAEW).

Tarnveer Singh is Director (Security and Compliance) at Cyber Wisdom Ltd.. He is an experienced Chief Information Security Officer (CISO) in the insurance sector and many other sectors. He is a Fellow of the British Computer Society and Chartered Institute of Information Security.

Finance Transformation
Leadership on Digital Transformation and Disruptive Innovation

Aikta Varma and Tarnveer Singh

CRC Press
Taylor & Francis Group
Boca Raton London New York

CRC Press is an imprint of the
Taylor & Francis Group, an **informa** business

Designed cover image: Shutterstock Images

First edition published 2025
by CRC Press
2385 NW Executive Center Drive, Suite 320, Boca Raton FL 33431

and by CRC Press
4 Park Square, Milton Park, Abingdon, Oxon, OX14 4RN

CRC Press is an imprint of Taylor & Francis Group, LLC

ISBN: 9781032846897 (hbk)
ISBN: 9781032844190 (pbk)
ISBN: 9781003514503 (ebk)

DOI: 10.1201/9781003514503

Typeset in Minion
by KnowledgeWorks Global Ltd.

Contents

CHAPTER 1

The Context and Case for Finance Transformation

WHILE COMPLETING MY EXECUTIVE MBA at Warwick University, I accomplished research interviewing senior finance leaders across the financial services sector. What I soon realised was that these conversations would help me identify deep-rooted problems with how finance is approached and lead to further research. My research with industry leaders set in motion a desire to uncover potential solutions and share best practices to help colleagues address these fundamental issues. My research helped me to consider how we can genuinely transform our finance functions to be significantly more effective.

Accounting, as a profession is as old as recorded language itself, has witnessed significant transformations throughout history. With the advent of technology, the field of accounting has embraced automation, changing the way financial data is processed, analysed, and interpreted. In this book, we will explore the origins of accounting through its modern-day digital transformation. We will also delve into the future of accounting, examining the role of cloud computing, machine learning (ML), and other innovative technologies in shaping the accounting profession.

Accounting, arguably one of the oldest professions in human history, traces its roots back more than five thousand years. The earliest known bookkeeping records date to ancient Egypt, where King Scorpion I's tomb contained ivory tablets etched with hieroglyphs, indicating financial transactions and tributes from various towns. These records demonstrate that financial record-keeping is inexorably linked with the written word and the need to document trade and business transactions.

As civilisations progressed, banking records were discovered in ancient Greek and Roman cities, providing evidence of early accounting practices. However, it wasn't until the European Renaissance of the 15th and 16th centuries that modern accounting was born.

Italian mathematician Luca Pacioli, often regarded as the father of modern accounting, documented the process of double-entry bookkeeping in his book, 'Summa de Arithmetica, Geometria, Proportioni et Proportionalita'. This revolutionary accounting methodology established the use of debits and credits in every transaction, forming the foundation of modern accounting systems.

The evolution of technology has played a crucial role in shaping the accounting profession. In the late 19th century, as commerce grew rapidly, the need for professional accountants expanded. The introduction of adding machines, such as William Burroughs' invention, revolutionised calculations, enabling accountants to perform their tasks with increased speed and efficiency. Subsequently, the advent of punch card machines in the early 20th century further enhanced data processing capabilities.

The true transformation of accounting came with the emergence of computer technology. In the 1950s, General Electric purchased the first computer specifically designed to run payroll, replacing the labour-intensive punch card method. With the introduction of personal computers in the 1980s, accounting processes became more accessible to businesses of all sizes. The late 1990s witnessed the launch of QuickBooks, a popular accounting software still widely used today.

Cloud computing has emerged as a game-changer for the accounting profession, allowing financial records to be stored securely and accessed remotely. The cloud-based approach eliminated the need for physical filing cabinets and simplified data management.

Consequently, accounting professionals gained access to real-time data, tighter security, and improved data retention.

The advent of cloud computing and the democratisation of data have significantly transformed the accounting landscape. Previously, accounting information was accessible only to a select few within the finance team, leading to inefficiencies and information bottlenecks. However, with cloud-based solutions, accounting professionals gained wider access to data, enabling them to make informed decisions and provide valuable insights to their organisations.

Cloud computing not only improved data accessibility but also facilitated automation in accounting processes. Manual data entry tasks, such as invoice processing and bank account reconciliation, were streamlined through the use of ML algorithms. Optical Character Recognition and Intelligent Data Capture technologies allowed businesses to scan and digitise invoices, reducing the need for manual data entry. As a result, accountants could focus on higher-level analysis and value-added activities, leading to increased accuracy, speed, and efficiency.

ML, a subset of *artificial intelligence (AI),* has emerged as a powerful tool in the accounting profession. ML algorithms can process and analyse vast amounts of financial data, identifying patterns, anomalies, and correlations that may not be apparent to human analysts. By training ML tools to interpret invoices, emails, and other financial documents, accountants can automate data extraction and analysis, reducing the time spent on manual tasks.

ML algorithms not only streamline accounting processes but also enable better decision-making. By continuously learning from financial data, these algorithms can provide valuable insights and predictions, helping organisations optimise their financial strategies, identify cost-saving opportunities, and mitigate risks. ML-driven automation empowers accountants to focus on strategic financial planning, analysis, and decision-making, maximising their value to their organisations.

In my experience, digital transformation in accounting offers numerous benefits to organisations of all sizes. By embracing cloud computing, ML, and other innovative technologies, businesses can:

Improve efficiency: Automation of manual tasks saves time and reduces human error, enabling accountants to focus on higher-value activities.

Enhance accuracy: ML algorithms can process large volumes of data accurately and quickly, minimising the risk of human error.

Streamline data management: Cloud-based solutions facilitate seamless data storage, retrieval, and collaboration, promoting efficient workflows.

Enable real-time insights: Access to real-time financial data empowers accountants to make informed decisions and respond promptly to market changes.

Drive strategic decision-making: ML algorithms provide data-driven insights and predictions, facilitating better financial planning and risk management.

Increase cost savings: Automation reduces the need for manual labour, leading to cost savings and improved operational efficiency.

Foster innovation: The adoption of innovative technologies encourages a culture of innovation within the accounting profession, pushing the boundaries of what's possible.

Thottoli et al. (2023) contend that the future of accounting lies in embracing ongoing digital transformation. This is based on an analysis of 288 peer-reviewed research articles and the assessment of several cutting-edge digital technologies, which are looked in greater depth in this book.

Cloud computing, ML, and other emerging technologies will continue to shape the accounting profession, increasing efficiency, accuracy, and strategic decision-making capabilities. As accountants and finance professionals adapt to these changes, they will unlock new opportunities for growth and innovation, propelling the field into the era of deep finance.

In my experience, the journey towards deep finance requires a shift in mindset and a willingness to embrace technological advancements. Accountants must be open to learning new skills, such as data analysis and ML, to stay relevant in a rapidly changing business environment. By leveraging the power of technology, accountants can elevate their roles, provide greater value to their organisations, and drive the future of accounting forward.

As we embark on this transformative journey, it is essential to remember that technology is a tool, not a replacement for human expertise. Accountants will continue to play a crucial role in analysing, interpreting, and communicating financial information. By combining their domain knowledge with the power of technology, accountants can shape the future of accounting and contribute to the success of their organisations.

The history and future of accounting are intertwined with technological advancements. From the ancient origins of bookkeeping to the digital transformation fuelled by cloud computing and ML, accounting has evolved to meet the changing needs of businesses.

Embracing this digital transformation offers numerous benefits, from increased efficiency and accuracy to improved decision-making and cost savings.

As accounting professionals navigate this journey towards *deep finance*, they must embrace technology, stay updated with the latest advancements, and continuously enhance their skills. By doing so, accountants can unlock new opportunities, provide greater value to their organisations, and shape the future of accounting in the digital age. The future is now, and the possibilities for accounting are limitless.

Finance Transformation is increasingly important in today's fast-paced and ever-evolving business landscape, which has gained significant prominence. Finance transformation refers to the strategic initiative undertaken by organisations to enhance their finance function and align it with their overall business objectives. It involves reimagining and restructuring financial processes, systems, and technologies to drive efficiency, improve decision-making, and achieve sustainable growth.

One of the key elements of finance transformation is the shift from traditional finance models to more agile and data-driven approaches. This entails leveraging advanced analytics, automation, and AI to streamline financial operations, optimise resource allocation, and uncover valuable insights. By embracing finance transformation, companies can unlock their full potential, enabling them to adapt to changing market dynamics, mitigate risks, and capitalise on emerging opportunities.

In recent years, several trends have emerged that are shaping the landscape of finance transformation. One such trend is the increasing adoption of cloud-based financial management systems. By migrating their finance operations to the cloud, organisations can benefit from enhanced scalability, improved security, and real-time access to critical financial data. This allows for more accurate and timely financial reporting, better collaboration among teams, and greater flexibility in adapting to changing business needs.

Another significant trend in finance transformation is the integration of ML and predictive analytics into financial processes. By leveraging these technologies, companies can automate routine tasks, such as data entry and reconciliation, while also gaining valuable insights into future financial performance. ML algorithms can analyse vast amounts of historical financial data to identify patterns and trends, enabling organisations to make more informed and strategic decisions.

Additionally, there is a growing focus on sustainability and environmental, social, and governance (ESG) factors in finance transformation. Companies are recognising the importance of integrating sustainability considerations into their financial strategies and operations. This involves measuring and reporting on ESG performance, incorporating ESG criteria into investment decisions, and aligning financial goals with broader sustainability objectives.

Technology has been the main driving force behind the finance transformation revolution. The advancements in digital technologies have enabled organisations to automate manual processes, improve accuracy, and enhance decision-making capabilities. One of the key technologies that have revolutionised finance transformation is robotic process automation (RPA). RPA involves the use of software robots to automate repetitive and rule-based tasks, such as invoice processing and accounts payable/receivable. By offloading these tasks to robots, finance teams can focus on more strategic activities, such as financial analysis and risk management.

Another technology that has had a profound impact on finance transformation is blockchain. Blockchain technology provides a decentralised and secure platform for recording and verifying financial transactions. It eliminates the need for intermediaries, reduces the risk of fraud, and enhances transparency. By leveraging blockchain, organisations can streamline processes, such as trade finance and supply chain financing, while also improving the efficiency and security of cross-border transactions.

AI and ML are also playing a crucial role in finance transformation. AI-powered chatbots and virtual assistants are being used to automate customer interactions, provide personalised financial advice, and enhance customer experience. ML algorithms are being deployed to analyse vast amounts of financial data and identify patterns that can help in forecasting and risk management. These technologies are revolutionising the finance industry by enabling faster, more accurate, and data-driven decision-making.

The adoption of finance transformation offers numerous benefits to organisations across industries. One of the key benefits is improved operational efficiency. By automating repetitive tasks and streamlining processes, companies can reduce costs, eliminate errors, and enhance productivity. This allows finance teams to focus on value-added activities such as financial analysis, forecasting, and strategic planning.

Finance transformation also enables organisations to make more informed and timely decisions. By leveraging advanced analytics and real-time data, companies can gain valuable insights into their financial performance, market trends, and customer behaviour. This empowers finance leaders to proactively identify risks, capitalise on opportunities, and drive business growth.

Another significant benefit of finance transformation is enhanced compliance and risk management. With the increasing regulatory scrutiny in the financial industry, organisations need to ensure that their financial processes and reporting are accurate, transparent, and compliant. By implementing robust financial management systems and controls, companies can minimise the risk of non-compliance and mitigate financial and reputational risks.

Furthermore, finance transformation can lead to improved stakeholder engagement. By providing stakeholders with accurate and timely financial information, organisations can build trust, enhance transparency, and strengthen relationships with investors, customers, and business partners. This, in turn, can attract new investment opportunities, foster growth, and create a competitive advantage.

Regulatory challenges are an important consideration. While digital transformation in the financial services industry offers numerous benefits, it also presents regulatory challenges that organisations need to overcome. One of the key challenges is ensuring data privacy and security. With the increasing volume and complexity of financial data, organisations must ensure that customer information is protected from unauthorised access and cyber threats.

Compliance with data protection regulations, such as the General Data Protection Regulation, is critical to maintaining customer trust and avoiding hefty fines.

Another regulatory challenge is managing the risks associated with emerging technologies. As organisations adopt technologies like blockchain and AI, they need to address issues related to data accuracy, validation, and governance. Regulators are closely monitoring the use of these technologies to ensure that they do not compromise market integrity or pose systemic risks. Organisations need to establish robust controls and frameworks to manage these risks and comply with regulatory requirements.

Furthermore, regulatory compliance itself can be a challenge in the context of finance transformation. Financial institutions are subject to a myriad of regulations, such as Basel III and International Financial Reporting Standards. Implementing finance transformation initiatives while ensuring compliance with these regulations requires careful planning, coordination, and monitoring. Organisations need to work closely with regulators, invest in robust compliance frameworks, and conduct regular audits to demonstrate adherence to regulatory requirements.

As we look to the future, the pace of finance transformation is expected to accelerate even further. Advancements in technologies like AI, blockchain, and data analytics will continue to reshape the finance industry, enabling organisations to unlock new opportunities and navigate complex challenges. The integration of finance and sustainability will become increasingly important as organisations recognise the need to align financial goals with broader environmental and social objectives.

Furthermore, the role of finance professionals will undergo a significant transformation. With the automation of routine tasks, finance professionals will be able to focus on more strategic activities, such as financial analysis, risk management, and decision support. They will need to develop new skills, such as data analytics, strategic thinking, and leadership, to thrive in the digital finance landscape.

Finance transformation is revolutionising the industry by enabling organisations to enhance their operational efficiency, improve decision-making, and achieve sustainable growth. By leveraging advanced technologies and embracing data-driven approaches, companies can unlock their full potential and thrive in today's dynamic business environment. With the future of finance transformation promising even greater advancements, organisations need to embrace change, invest in innovation, and seize the opportunities that lie ahead.

CHAPTER 2

Finance Transformation

*Revolutionising the Future
of Financial Management*

F INANCE TRANSFORMATION HAS BECOME a crucial business strategy that is gaining traction across industries worldwide. In today's fast-paced business landscape, companies must continually evolve to stay competitive. As finance professionals, we must adapt to this new reality by embracing financial transformation. This book will delve into the exciting world of finance transformation and explore how businesses can harness their power to take their financial management practices to the next level.

Finance transformation is a strategic initiative that aims to revamp financial systems, processes, and capabilities to enhance business performance and drive sustainable growth. It goes beyond just implementing new software and tools; it involves a holistic approach that encompasses everything from financial planning and analysis (FP&A) to accounting, reporting, and risk management.

Papathomas et al. (2023) argue that the key enablers to digital transformation within finance are strategy and organisation, people and culture, technology and innovation, and lastly, value proposition (Papathomas and Konteos, 2023). Based on my own experience, I concur that these are contingency factors that enable digital change, but there are other factors that impact change that I will refer to in later chapters.

In my experience, the main drivers of finance transformation are to streamline financial operations, improve decision-making, and enhance the efficiency and effectiveness of financial processes. By understanding the goals, customer needs, and market dynamics of the company, businesses can create a roadmap that aligns with their overall strategy and drives transformation.

Finance transformation is essential in today's digital age. Chief Finance Officer (CFO) have a critical role to play in enabling their finance function to deliver faster and more accurate data to support critical decision-making processes. Failure to embrace digital transformation can result in falling behind competitors and missing out on valuable opportunities.

The benefits of finance transformation are numerous. Firstly, it provides real-time access to financial data across the organisation, allowing for better identification of potential issues and opportunities, monitoring of financial performance, and measurement of progress against key performance indicators (KPIs). This comprehensive view of financial data improves communication with stakeholders, builds trust, and enhances the organisation's reputation.

Secondly, finance transformation leads to more streamlined business processes and financial operations. By eliminating manual errors and improving accuracy, finance teams can save time and resources. This streamlined approach also creates a better experience for customers and suppliers, accelerating the order-to-cash cycle and eliminating bottlenecks in the financial process.

Thirdly, finance transformation establishes a single source of truth for financial data, reducing confusion and errors. All stakeholders have access to the same data, eliminating discrepancies and manual reconciliations. This results in more accurate and reliable decision-making and reduces the risk of data breaches or other security issues.

Automation plays a crucial role in finance transformation. By automating time-consuming, manual finance tasks such as invoice processing and account reconciliation, finance teams can improve accuracy, reduce the risk of errors, and free up time for more strategic activities.

Lastly, finance transformation improves collaboration and communication across the organisation. By providing better access to financial data and streamlining financial processes, organisations can enhance collaboration, achieve strategic goals, and gain a competitive advantage.

In today's rapidly evolving business landscape, finance transformation has become a necessity rather than an option. The finance function plays a crucial role in enabling the strategic ambitions of the enterprise, and CFOs are under constant pressure to do more, reduce costs, and innovate. A significant percentage of C-level executives in accounting and finance expect the development of intelligent, automated accounting systems to have the highest impact in the next three years.

Finance transformation initiatives are key to maximising business profitability. By evaluating, prioritising, and scoping activities in a finance transformation roadmap, organisations can focus on the differentiators of success. Creating a future-state design with best practices, case studies, and tools strengthens the business case for finance transformation.

Implementing strategies that create value for business partners is crucial. The finance function must not only drive enterprise digital initiatives but also digitalise itself. By leveraging digital technologies, finance can streamline processes, improve efficiency, and generate actionable business insights.

The *Finance Operating Model* is vital. To drive more business impact, finance functions need to refocus and restructure their operating model, processes, and capabilities. By adopting more sophisticated finance analytics and IT solutions, finance can enhance its ability to generate valuable insights and support informed decision-making. This shift towards a data-driven finance function is crucial for driving good decisions and improving performance.

Benchmarking metrics that are key to finance operational efficiency is a great starting point for finance transformation efforts. By identifying areas that need improvement, organisations can drive productivity, cut costs, and build a more responsive finance function.

Analytics play an increasingly critical role in finance transformation. Progressive finance leaders stay ahead of data and analytics innovations to build a more data-driven finance function. By embracing these technologies, finance can generate actionable business insights that drive performance improvement.

Understanding the latest trends and innovations in data and analytics is essential for finance leaders. By staying informed, finance functions can leverage the power of analytics to unlock valuable insights and drive strategic decision-making.

Talent development is crucial to finance transformation. The shifting finance talent landscape necessitates equipping the finance function with next-generation skills. Finance leaders must optimise and lead the finance team of the future. By nurturing the right talent and developing the necessary capabilities, organisations can ensure the success of finance transformation initiatives.

Leading the next-generation finance workforce requires a deep understanding of the skills and competencies needed. Finance leaders must adapt to the changing landscape, embrace new technologies, and foster a culture of continuous learning and development.

Business partnership activities are critical to finance transformation. Knowing when and how to use business partners like *shared services* is essential. By benchmarking and measuring success, organisations can maximise the value of shared services and drive effective finance transformation.

Understanding the current trends and best practices impacting shared services, strategy and structure is crucial. By staying informed, finance leaders can make informed decisions on how to leverage shared services to streamline processes and optimise performance.

Finance transformation can yield significant benefits for organisations. By taking action now and avoiding common mistakes, companies can gain a competitive edge, drive business results, and prepare for the future. Future-ready companies embrace change, leverage new technologies, and continuously strive for improvement.

Finance transformation is no longer an option but a necessity for businesses striving to maximise profitability. By evaluating, prioritising, and implementing finance transformation initiatives, organisations can drive business impact, improve operational efficiency, and generate actionable insights. By embracing digital technologies, analytics, and developing the right talent, finance can position itself as a strategic partner for the business.

Avira et al. (2023) make the case that in order to succeed in digital transformation in financial management, companies must adopt a strategic approach, mitigate risks, and engage stakeholders effectively (Avira et al., 2023).

In my own experience, I understand how embarking on a finance transformation journey comes with its fair share of challenges. It is important to address these challenges to ensure a successful transformation. Some common challenges include resistance to change, integration issues, data quality issues, lack of expertise, budget constraints, time constraints, lack of a digital transformation strategy, and security concerns.

Resistance to change is a common challenge when implementing finance transformation. Finance professionals may be accustomed to traditional ways of working and may be reluctant to embrace new methods. However, by involving employees in the transformation process, communicating the benefits of the transformation, and providing training and support, organisations can overcome this challenge and get their finance teams on board with the change.

Integration issues can arise when trying to integrate new systems and processes with existing ones. It is important to assess the compatibility of new systems and conduct pilot testing to identify and address integration issues before they become major problems.

Data quality is critical for successful finance transformation. Establishing clear data governance policies and procedures, conducting regular data quality checks and audits, and using the right tools and processes can help maintain data quality and mitigate the risk of data-related issues.

Lack of expertise in finance transformation can be a challenge for organisations. Hiring external consultants or partnering with vendors who specialise in finance transformation can help ensure the effective and efficient implementation of the transformation.

Budget constraints can limit the resources available for finance transformation. Exploring different financing options and prioritising investments based on expected ROI can help organisations make the most of their resources.

Time constraints can be a challenge when balancing the need for transformation with day-to-day operations. Creating a realistic timeline, allocating sufficient resources, and breaking the project down into smaller, manageable components can help organisations stay on track and meet their transformation goals.

Lack of a clear digital transformation strategy can lead to a lack of focus and wasted resources. Developing a comprehensive strategy that aligns with overall business goals and includes a roadmap for finance transformation can provide direction and ensure that efforts are aligned with desired outcomes.

Security concerns are a crucial consideration when implementing new financial systems and processes. Implementing robust security measures and protocols, such as encryption, access controls, and regular security audits, can help protect sensitive financial data and mitigate security risks.

Automation and digital transformation play a significant role in finance transformation. By leveraging automation and machine learning technologies, finance teams can make better-informed decisions, reduce manual workloads, and increase efficiency.

Automation eliminates time-consuming and error-prone manual processes by automating tasks like accounts payable and receivable. This not only saves time and resources but also improves accuracy. By freeing up time, finance teams can focus on more strategic tasks such as strategic planning, financial analysis, and risk management.

Machine learning takes automation a step further by analysing historical financial data to identify patterns and trends. These insights enable more accurate predictions about future financial performance and help identify potential financial risks. By leveraging machine learning, CFOs can improve financial forecasting and risk management.

While automation and machine learning technologies offer significant benefits, it is important to note that they are not a replacement for human expertise. CFOs must exercise sound judgement and use these tools to augment their financial expertise rather than replace it.

A *Finance Transformation Roadmap* must be built. Building a roadmap is crucial for a successful transformation. It involves assessing the current state, designing the future state, building the new processes and systems, and operating the transformed finance function.

To assess the current state, organisations should identify pain points and areas of opportunity for improvement within their finance function. This includes evaluating current financial systems and processes, as well as the skills and capabilities of the finance team. Conducting a gap analysis can help compare the current state to the desired future state and define the key initiatives for transformation.

In the design stage, organisations should create a blueprint for the future state of the finance function. This includes defining new financial processes, selecting new financial systems, and defining new roles and responsibilities for the finance team. Best practices, benchmarking against industry standards, and alignment with the overall business strategy should guide the design process.

Once the blueprint is in place, organisations can start building the new processes and systems. This involves configuring the new financial systems, migrating data from old systems to new ones, and training the finance team on the new processes and systems. Communication with the broader organisation and preparation for potential impacts on other departments should also be considered during this stage.

Operating the transformed finance function involves running the new financial processes and systems, monitoring performance, and continuously improving the function. Establishing performance metrics, monitoring progress, making adjustments as needed, and celebrating successes are key aspects of this stage. Continuous improvement ensures that the finance function remains efficient and effective in supporting the organisation's financial goals.

By following these steps and developing a well-designed roadmap, organisations can achieve a more efficient and effective finance function that drives better financial performance and supports strategic goals.

Finance transformation is a critical process that enables organisations to improve their financial performance and achieve strategic goals. By embracing automation, machine learning, and digital transformation, CFOs can make better-informed decisions, reduce manual workloads, and increase efficiency. While finance transformation comes with its challenges, organisations can overcome them by addressing resistance to change, integrating new systems effectively, ensuring data quality, acquiring the necessary expertise, managing budgets and time constraints, developing a clear digital transformation strategy, and prioritising security.

Building a finance transformation roadmap is essential for a successful transformation. By assessing the current state, designing the future state, building the new processes and systems, and operating the transformed finance function, organisations can achieve a more streamlined and effective finance function that drives better financial performance and supports the organisation's strategic goals.

Finance transformation is not just a buzzword; it is a strategic imperative in today's digital age. By embracing finance transformation, organisations can revolutionise their financial management practices and position themselves for sustainable growth in a rapidly changing business landscape.

Finance transformation is a critical process for organisations seeking to adapt and thrive in a rapidly changing business landscape. As finance leaders grapple with the challenges of modernisation, they must consider various factors that drive the need for change, the role of leadership in guiding the transformation, and the best approach to implementation. CFOs must explore the key elements of finance transformation, drawing insights from industry experts and thought leaders.

Catalysts for change must be understood. Finance delivery change can be triggered by various catalysts, ranging from a new vision created by the C-suite to external factors that leave organisations with no choice but to change. The single biggest success factor is having a CFO or business leader who can effectively communicate the need for change and paint a compelling picture of the future. This vision, coupled with the support of the broader leadership team, sets the stage for successful finance transformation.

In some cases, the movement towards a new delivery model is driven by external factors such as cost reduction, process efficiency drives, or the need to become more agile in response to changing market conditions. The author cites the example of an airline client that underwent a massive transformation due to the rise of online booking and low-cost competitors. The urgency of the situation forced them to quickly adopt shared services and outsourcing, ultimately reducing headcount and automating finance transaction processing.

Peer pressure can also play a role in stimulating change. When competitors or industry leaders adopt a new finance delivery model, it becomes easier to convince stakeholders of the need for transformation. Nevertheless, there are clear implications of running with the herd and the influence it can have on driving change within the finance function.

However, finance leaders also recognise the importance of being proactive rather than reactive. Finance leaders should steer the vision and strategy of finance rather than waiting for organisational indicators or external pressures. Creating discontent with the status quo can serve as a scare factor and motivate the finance function to embrace change, even in the absence of a burning platform.

Leadership in driving change is vital. While the CFO is typically the sponsor of any finance delivery transformation, the Chief Executive Officer (CEO) involvement and support are crucial for success. The CEO's visible sponsorship gives validity to the functional change and ensures alignment with the overall business strategy. However, the extent of the CEO's involvement may vary depending on factors such as the scale of the company, the degree of change, and the state of the enterprise resource planning (ERP) platform.

It is important to emphasise the importance of the CEO's mandate in driving change within the finance function. Without the CEO's support, resistance from cynics within the organisation can hinder the progress of the transformation. The CEO's role in change management is dependent on the model, size of the company, and implications of the change. While senior management commitment is crucial, the level of active participation may vary depending on the scope of the transformation.

The CFO should act as the functional sponsor and create a burning platform for finance transformation. The CFO's role is particularly significant when the transformation involves repositioning the finance team as a valued business partner and strategic decision support team. However, if the transformation primarily focuses on transaction processing, the CEO's sponsorship may be less important.

When it comes to the velocity and extent of change, finance leaders approach finance transformation in different ways. Some advocate for a *revolution*, urging finance leaders to 'get on with it' and create momentum by moving quickly. A rapid revolution builds momentum and allows for continuous improvement in the long term.

On the other hand, companies can no longer afford to take matters slowly and must embrace a more *evolutionary* approach. We can highlight how a big change naturally results in an evolutionary approach, given the number of stakeholders involved. There is also a need to emphasise the importance of multiple phases within finance transformation, with revolutions establishing the foundation for continuous improvement and evolution.

However, the speed of change must be carefully planned and aligned with the organisation's resources and capabilities. It is crucial to highlight the importance of considering the organisation's size and scale when determining the speed of change. Smaller organisations may need to deliver small, quick wins to build credibility when resources are limited. Planning and finding the right starting point for finance transformation are critical.

The *Deployment Plan* must be optimised. Deploying shared services, or outsourcing, in the finance function requires careful planning and consideration of various factors. The optimal approach depends on the maturity of processes, the alignment with the corporate centre, and the potential resistance to change from different geographies or business units within the organisation.

Finance leaders generally prefer a process-primacy approach, focusing on the maturity of processes rather than geography or business unit deployment. We can highlight the importance of starting with the least mature process and gradually expanding to other areas. It may be useful at this point to suggest a 'design by process and implement by geography' approach, starting with areas that can deliver early wins and demonstrate the effectiveness of shared services or outsourcing.

The decision to 'lift and shift' or 'fix and shift' by process and geography plays a crucial role in deployment. Culture also factors into the decision, with some organisations being more inclined to manage supply chains and others having an aversion to outsourcing due to a 'not invented here' mentality. Scale and economics should also be considered, as finance departments must carefully evaluate the cost-effectiveness of outsourcing.

We can also highlight the importance of sequencing deployment based on cultural readiness. It is worth noting that organisations should start with regions or business units that are more accepting of shared services or outsourcing and gradually extend the transformation to regions that may be more resistant to change.

The *Sourcing Strategy* must be right. The decision to make or buy, whether to pursue shared services or outsourcing, depends on various factors such as the organisation's vision for the finance function, scale, need for expertise, process maturity, industry sector, and risk tolerance. The organisation's vision plays a crucial role in determining the most suitable model. It is worth emphasising the importance of designing by process and implementing by geography, starting in areas that can deliver early wins and gradually scaling up. The need for expertise is another critical factor. Outsourcing providers often bring specialised knowledge and capabilities that exceed those of their clients. If upskilling or access to specific finance capabilities is essential, outsourcing may be the preferred option. The importance of considering both critical mass and organisational maturity is vital when determining the sourcing strategy.

The organisation's *current operations* and footprint can also influence the decision. Organisations with existing operations in Eastern Europe may find it cost-effective to leverage their footprint and build a shared services centre. However, developing a captive operation in a completely new location, such as India, may be less favourable compared to outsourcing.

Corporate culture also plays a significant role in the decision-making process. Some finance organisations prefer to deliver services internally and have a solutions-based culture, while others have a strong aversion to outsourcing. The organisation's risk appetite should also be assessed, as risk tolerance can vary depending on the organisation's strategic objectives and industry sector.

Overcoming *internal change battles* is important. While executive leaders may endorse the need for finance transformation, the hardest change battles are often fought within the finance function itself. Traditional accountants, accustomed to rules and routines, may resist the shift towards shared services or outsourcing. As processes and responsibilities change, finance professionals must adapt to new roles and technologies, which can be challenging, especially for those with long tenures and a focus on compliance.

It is worth emphasising the importance of understanding what needs to stay the same and when behaviours need to change. While certain aspects of finance operations, such as integrity and stewardship, should not be compromised, other aspects, such as bureaucracy and over-engineering of processes, may need to be re-evaluated.

Communication plays a vital role in convincing the finance team of the need for change. Failure to effectively communicate the rationale and vision of the transformation can hinder progress. Finance leaders must invest in new career models, highlight the business imperative for change, and provide ongoing support to help the finance team navigate the transformation.

Ring-fencing the change and managing it effectively is crucial for successful finance transformation. Change management programmes should be in place to guide employees through the transition and ensure that the middle layer of finance management is retained. It is worth noting the importance of maintaining momentum and accepting change within the finance function.

Leadership has an important role to play. The success of finance transformation depends greatly on effective leadership. Whether an internal or external hire, the leader must possess the necessary institutional knowledge and bring fresh thinking or proven methodologies to the table. The decision to hire internally or externally depends on the context in which the leader will operate.

While institutional knowledge is important, we must recognise the value of a fresh perspective. External leaders can bring objectivity and compare the organisation with others, providing valuable insights and benchmarking. However, the success of an external leader may depend on their ability to foster strong relationships with senior management and finance reports.

External resources, such as consultants, can provide a structured approach to change management and offer a defined formula and sequence of events. Their experience and expertise can enhance the change process and ensure efficiency. It is worth mentioning the important use of external perspectives, highlighting the challenge organisations face in looking authentically and self-critically at themselves.

Outsourcing providers can also play a role in change management, leveraging their breadth of experience and practicality to drive successful transformation. The partnership between external and internal resources is crucial, with external perspectives helping navigate the organisation and ensure effective change management.

Finance transformation is a complex process that requires careful planning, a clear vision, effective communication, and strong leadership. The catalysts for change can range from internal visions created by the C-suite to external factors that leave organisations with no choice but to transform. The role of leadership, particularly the CEO and CFO, is critical in driving and supporting change within the finance function.

The speed and extent of change can vary, with some organisations opting for a revolution while others embrace a more evolutionary approach. The deployment plan should consider process maturity, geography, and the organisation's readiness for change. The decision to make or buy, whether to pursue shared services or outsourcing, depends on various factors, including the organisation's vision, scale, expertise, and risk tolerance.

Overcoming internal change battles within the finance function is crucial for successful transformation. Effective communication, change management programmes, and ongoing support are essential to help finance professionals adapt to new roles and technologies. Effective leadership, whether internal or external, plays a significant role in driving and guiding the transformation process.

Finance transformation is a continuous journey, requiring patience, flexibility, and an alignment between capability and ambition. By navigating the path to change with careful planning, effective leadership, and a focus on the organisation's vision, finance leaders can successfully transform their functions and position their organisations for long-term success.

Measuring the success of finance transformation is crucial to understand the impact and effectiveness of the implemented changes. *KPIs* play a vital role in evaluating the success of finance transformation initiatives. Let's explore some key metrics that organisations can use to measure the success of their finance transformations.

Cost savings and operational efficiency: One of the primary goals of finance transformation is to improve operational efficiency and reduce costs. Organisations can measure the success of their transformation by analysing cost savings achieved through process automation, resource optimisation, and streamlined workflows.

Improved financial reporting and analysis: A successful finance transformation should result in improved financial reporting capabilities. Organisations can assess the success of their transformation by evaluating the timeliness, accuracy, and comprehensiveness of financial reports. Enhanced data analytics capabilities and real-time insights also contribute to measuring the effectiveness of finance transformations.

Enhanced decision-making capabilities: Finance transformations should empower organisations to make data-driven decisions. Measuring the success of transformation initiatives can be done by evaluating the quality of decision-making, speed of decision-making, and the ability to respond to market changes effectively.

Employee satisfaction and engagement: Finance transformations impact the entire organisation, and employee satisfaction and engagement are crucial for long-term success. Organisations can measure the success of their transformations by conducting employee surveys, assessing change management effectiveness, and monitoring employee feedback.

CHAPTER 3

Finance Strategy

Finance strategy is a crucial aspect of any organisation, regardless of its size or industry. It lays the foundation for the financial stability and success of a business. A well-defined finance strategy provides direction, guidance, and a roadmap for achieving growth and profitability. It encompasses various crucial elements, such as financial planning, budgeting, risk management, and investment strategies. By understanding the importance of finance strategy, businesses can unlock the secrets to sustainable growth and long-term profitability.

In my experience, a robust finance strategy ensures that an organisation's financial resources are allocated effectively and efficiently. It enables businesses to make informed decisions about investments, cost management, and resource allocation. Without a clear finance strategy, organisations may find themselves facing financial instability, missed opportunities, and unnecessary risks. Therefore, it is essential to recognise the significance of finance strategy and its impact on the overall success of a business.

An effective *finance strategy* comprises several key components that work together to drive growth and profitability. These components include financial planning, risk management, performance measurement, and capital allocation.

Financial planning is the foundation of a successful finance strategy. It involves setting financial goals, forecasting revenues and expenses, and developing a budget. A well-designed financial plan provides a roadmap for achieving the desired financial outcomes. It helps businesses identify potential risks and opportunities, enabling them to make informed decisions about resource allocation and investment priorities.

Risk management is an integral part of finance strategy. It involves identifying and mitigating potential risks that can impact an organisation's financial stability and performance. By implementing effective risk management practices, businesses can protect themselves from financial uncertainties and unexpected events. This includes assessing and managing market risks, credit risks, operational risks, and regulatory risks. A robust risk management framework ensures that businesses can navigate challenges while minimising the negative impact on their financial health.

DOI: 10.1201/9781003514503-3

Performance measurement is vital for evaluating the effectiveness of a finance strategy. It involves tracking key financial metrics, such as revenue growth, profitability, return on investment, and cash flow. By continuously monitoring and analysing these metrics, businesses can gain insights into their financial performance and identify areas for improvement. Performance measurement provides valuable feedback on the effectiveness of the finance strategy, enabling businesses to make data-driven decisions to optimise their financial outcomes.

Capital allocation refers to the process of distributing financial resources across different investment opportunities. An effective finance strategy includes a well-defined capital allocation framework that aligns with the organisation's goals and risk appetite. It involves evaluating investment opportunities, prioritising projects, and allocating resources based on their expected return on investment. A robust capital allocation strategy ensures that resources are allocated to projects with the highest potential for growth and profitability.

3.1 GROWTH

A finance strategy focused on growth aims to maximise the organisation's potential for expansion and market penetration. It involves identifying growth opportunities, developing investment strategies, and allocating resources strategically. Here are some key considerations for developing a finance strategy for growth:

Market Analysis: Conduct a thorough analysis of the market to identify growth opportunities and understand market dynamics. This includes analysing customer needs, the competitive landscape, and industry trends. By understanding the market, businesses can develop targeted strategies to capitalise on growth opportunities.

Investment Prioritisation: Prioritise investment opportunities based on their potential for growth and profitability. This involves evaluating projects based on their expected return on investment, market demand, and strategic fit. By focusing resources on high-potential projects, businesses can maximise their chances of achieving sustainable growth.

Financial Forecasting: Develop accurate financial forecasts to support growth initiatives. Financial forecasting involves projecting revenues, expenses, and cash flows based on historical data, market trends, and growth assumptions. These forecasts provide valuable insights into the financial implications of growth strategies and help businesses make informed decisions about resource allocation.

Capital Structure Optimisation: Optimise the organisation's capital structure to support growth objectives. This includes evaluating the mix of debt and equity financing, managing working capital effectively, and ensuring access to adequate funding sources. By optimising the capital structure, businesses can enhance their financial flexibility and capacity for growth.

3.2 PROFITABILITY

A finance strategy focused on profitability aims to maximise the organisation's financial performance and bottom line. It involves optimising revenue streams, cost management, and operational efficiency. Here are some key considerations for developing a finance strategy for profitability:

Revenue Optimisation: Identify opportunities to optimise revenue streams and enhance profitability. This includes analysing pricing strategies, exploring new markets, and developing innovative products or services. By maximising revenue generation, businesses can improve their overall profitability.

Cost Management: Implement effective cost management practices to control expenses and improve profitability. This involves analysing cost structures, identifying cost-saving opportunities, and optimising resource allocation. By reducing unnecessary costs and improving operational efficiency, businesses can enhance their profitability.

Working Capital Management: Optimise working capital management to improve cash flow and profitability. This includes managing inventory levels, optimising accounts receivable and payable, and minimising cash conversion cycles. By effectively managing working capital, businesses can unlock cash flow and improve their financial performance.

Operational Efficiency: Enhance operational efficiency to reduce costs and improve profitability. This involves streamlining processes, implementing automation and technology solutions, and continuous improvement initiatives. By improving operational efficiency, businesses can eliminate waste, reduce costs, and enhance profitability.

Implementing a finance strategy can be challenging due to various factors. Some common challenges businesses may face include:

Lack of Alignment: Lack of alignment between the finance strategy and the overall business strategy can hinder successful implementation. It is essential to ensure that the finance strategy is developed in alignment with the organisation's goals, objectives, and risk appetite.

Resistance to Change: Resistance to change can impede the implementation of a finance strategy. It is crucial to communicate the benefits of the strategy to stakeholders and proactively manage change to overcome resistance.

Limited Resources: Limited resources, such as financial, human, or technological, can pose challenges in implementing a finance strategy. It is important to allocate resources effectively and prioritise initiatives based on their potential impact.

Complexity: The complexity of financial processes, regulations, and systems can make finance strategy implementation challenging. It is crucial to simplify processes, ensure compliance, and leverage technology solutions to overcome complexity.

Lack of Data and Analytics: Inadequate data and analytics capabilities can hinder effective finance strategy implementation. It is important to invest in robust data management and analytics capabilities to support decision-making and performance measurement.

Technology plays a pivotal role in transforming finance strategy and driving growth and profitability. The advancements in technology have enabled businesses to automate financial processes, enhance data analytics capabilities, and improve decision-making. Here are some ways technology can support finance strategy transformation:

Automation: Technology enables the automation of repetitive financial processes, such as data entry, reconciliation, and reporting. Automation improves efficiency, reduces errors, and frees up resources to focus on value-added activities.

Data Analytics: Technology solutions provide powerful data analytics capabilities, enabling businesses to gain valuable insights into their financial performance. Advanced analytics tools can analyse large volumes of data, identify trends, and support data-driven decision- making.

Forecasting and Planning: Technology solutions facilitate accurate financial forecasting and planning. These tools leverage historical data, market trends, and sophisticated algorithms to generate reliable forecasts and support strategic decision-making.

Risk Management: Technology solutions enable businesses to enhance their risk management capabilities. Advanced risk management systems can assess and monitor risks in real-time, enabling businesses to proactively mitigate potential threats.

Collaboration and Communication: Technology tools facilitate collaboration and communication within finance teams and across the organisation. Cloud-based platforms, project management tools, and communication applications enable seamless collaboration, improving efficiency and decision-making.

Developing an effective finance strategy requires careful planning and consideration. Here are some best practices to guide the process:

Align with the Business Strategy: Ensure that the finance strategy aligns with the overall business strategy, goals, and objectives. This alignment ensures that financial resources are allocated strategically to support the organisation's growth and profitability.

Involve Key Stakeholders: Involve key stakeholders, such as senior management, finance teams, and other relevant departments, in the development of the finance strategy. This ensures buy-in, collaboration, and a holistic approach to strategy development.

Continuous Monitoring and Evaluation: Continuously monitor and evaluate the effectiveness of the finance strategy. Regularly review key financial metrics, performance indicators, and market trends to identify areas for improvement and make necessary adjustments.

Invest in Technology and Analytics: Invest in robust technology solutions and analytics capabilities to support finance strategy implementation. Leverage automation, data analytics, and forecasting tools to enhance efficiency, accuracy, and decision-making.

Risk Management Integration: Integrate risk management practices into the finance strategy. Consider potential risks and develop contingency plans to mitigate their impact on financial stability and performance.

There are case studies of successful finance strategy implementation we can consider, such as:

Volkswagen Group: Volkswagen Group, a global manufacturing company, successfully implemented a finance strategy focused on growth and profitability. They conducted a comprehensive market analysis, identifying emerging markets with high growth potential. By strategically allocating resources and investing in targeted expansion initiatives, they achieved significant revenue growth and market penetration.

Bank of America: Bank of America, a leading financial services firm, implemented a finance strategy aimed at improving profitability. They optimised their revenue streams by introducing new products and services and diversifying their customer base. By implementing cost-saving measures and enhancing operational efficiency, they achieved a substantial increase in profitability.

Finance Strategy Transformation Services can be a game-changer. To support businesses in unlocking the secrets of effective finance strategy, there are numerous finance strategy transformation services available. These services provide expertise and guidance in developing, implementing, and optimising finance strategies. They offer a range of solutions, including financial planning, risk management, performance measurement, and technology integration. By leveraging these services, businesses can accelerate their finance strategy transformation and achieve sustainable growth and profitability.

Understanding the importance of finance strategy is crucial for businesses seeking growth and profitability. By developing a well-defined finance strategy, focusing on key components, and leveraging technology, businesses can unlock the secrets to success.

Implementing best practices, overcoming common challenges, and learning from case studies can further enhance the effectiveness of finance strategy implementation. With the support of finance strategy transformation services, businesses can navigate the complexities of finance strategy and achieve their financial goals.

Finance transformation strategy has become a critical aspect for organisations to stay competitive and adapt to the evolving market dynamics. Developing a winning finance transformation strategy is essential for companies looking to optimise their financial processes, enhance efficiency, and drive sustainable growth. CFOs must consider the importance of finance transformation, key components of a winning strategy, and how to assess current finance processes. They must carefully set goals and objectives, develop a roadmap, implement the strategy, and measure success. Additionally, we will delve into real-life case studies showcasing successful finance transformation strategies.

Developing a finance transformation strategy is of paramount importance for organisations seeking to thrive in an increasingly complex and volatile business environment. By aligning financial processes with strategic objectives, companies can achieve operational excellence, enhance decision-making capabilities, and foster agility. A well-crafted finance transformation strategy enables organisations to streamline operations, reduce costs, and mitigate risks. Moreover, it empowers finance teams to focus on value-added activities such as financial analysis, planning, and forecasting rather than being bogged down by manual and repetitive tasks.

A winning finance transformation strategy comprises several key components that work in harmony to drive organisational success. Firstly, it is crucial to have a *clear vision* and *alignment* with overall business objectives. This ensures that the finance transformation strategy is in line with the organisation's strategic direction and fosters a unified approach across all departments. Secondly, *effective change management* is vital to ensure smooth implementation and adoption of the transformation strategy. This involves engaging stakeholders, communicating the benefits of the transformation, and providing the necessary training and support to employees.

Another critical component is *leveraging technology* and *automation*. Embracing digital solutions and tools can significantly enhance the efficiency and accuracy of financial processes, reduce manual errors, and improve data integrity. Furthermore, a winning finance transformation strategy emphasises the importance of data analytics and business intelligence. By harnessing the power of data, organisations can gain valuable insights, make informed decisions, and identify areas for improvement. Lastly, *continuous monitoring* and evaluation are essential to ensure the strategy remains aligned with the organisation's evolving needs and to identify opportunities for optimisation.

Before embarking on a finance transformation journey, it is crucial to conduct a thorough assessment of the current finance processes. This involves analysing existing workflows, identifying pain points, and understanding the root causes of inefficiencies or bottlenecks. One effective approach is to map out the end-to-end financial processes, from transactional activities to strategic reporting. This helps identify redundant or manual tasks that can be automated, streamline approval workflows, and eliminate unnecessary complexity.

Additionally, it is essential to gather feedback from key stakeholders, including finance staff, business partners, and customers. Their insights and perspectives can provide valuable input for identifying areas of improvement and aligning the finance transformation strategy with the organisation's strategic goals. By involving stakeholders early in the process, organisations can foster a sense of ownership and collaboration, leading to higher chances of successful implementation.

Setting *clear and measurable goals* is a fundamental step in developing a winning finance transformation strategy. These goals should be aligned with the organisation's strategic objectives and reflect the desired outcomes of the transformation. Examples of goals may include improving financial reporting accuracy, reducing the time required for closing the books, enhancing budgeting and forecasting capabilities, or increasing the efficiency of accounts payable and receivable processes.

To ensure the goals are achievable, it is essential to establish key performance indicators (KPIs) that can be tracked and monitored throughout the transformation journey. These KPIs could include metrics such as process cycle time, error rates, cost savings, or customer satisfaction levels. By defining clear goals and KPIs, organisations can measure the progress and success of the finance transformation strategy, while also providing a sense of direction and focus for the finance team.

In my experience, *roadmaps* are vital. Once the current finance processes have been assessed and goals and objectives have been established, the next step is to develop a roadmap for finance transformation. The roadmap serves as a strategic guide, outlining the key initiatives, activities, and timelines required to achieve the desired outcomes. It provides a structured approach for implementing the transformation strategy and ensures a cohesive and synchronised effort across different functions and departments.

The roadmap should include a detailed action plan, clearly defining the activities, responsibilities, and milestones for each phase of the transformation. It is essential to allocate resources effectively, both in terms of budget and personnel, to support the implementation of the roadmap. Regular checkpoints and progress reviews should be established to monitor the execution of the plan, identify any potential roadblocks, and make adjustments as necessary.

Implementing a finance transformation strategy requires careful planning and execution to ensure its successful adoption throughout the organisation. There are several key considerations to keep in mind during the implementation phase. Firstly, effective change management is critical. This involves communicating the benefits of the transformation to all stakeholders, addressing any concerns or resistance, and providing the necessary training and support to facilitate a smooth transition.

Secondly, strong leadership and sponsorship are essential. A dedicated project team, led by a senior executive, can provide the necessary direction, guidance, and support to drive the transformation forward. Additionally, it is crucial to foster a culture of continuous learning and improvement. This involves encouraging feedback, embracing innovation, and adapting to evolving technologies and best practices.

Measuring the success of finance transformation is crucial to evaluate the effectiveness of the strategy and identify areas for further improvement. KPIs established during the goal-setting phase play a vital role in measuring success. Regular monitoring and tracking of these KPIs enable organisations to assess the progress made, identify any gaps or bottlenecks, and make data-driven decisions to optimise the transformation strategy.

In addition to quantitative metrics, it is essential to gather qualitative feedback from stakeholders and end-users. This can be done through surveys, focus groups, or one-on-one interviews. By capturing the perspectives and experiences of those impacted by the transformation, organisations can gain valuable insights and identify opportunities for refinement or enhancement.

Case studies provide valuable insights into the practical implementation and impact of finance transformation strategies. Let's explore a few examples:

Unilever, a multinational manufacturing firm, embarked on a finance transformation journey to streamline its financial processes and enhance its decision-making capabilities. By leveraging advanced analytics tools, the firm was able to automate reporting and forecasting, resulting in a 30% reduction in reporting time and improved accuracy. The transformation also enabled the finance team to focus on strategic analysis, leading to better insights and recommendations for the organisation's growth.

Walmart, a retail giant, implemented a finance transformation strategy to optimise its accounts payable processes. By adopting an automated invoice processing system, it significantly reduced the invoice processing time, eliminated manual errors, and improved supplier relationships. The transformation resulted in cost savings of over 20% and enhanced cash flow management.

BNP Paribas, a financial services provider, undertook a finance transformation initiative to enhance its risk management capabilities. By implementing an integrated risk management system, the company was able to automate risk assessment, improve compliance, and strengthen decision-making processes. The transformation enabled them to proactively identify and mitigate risks, ensuring a more resilient and sustainable business.

These case studies highlight the tangible benefits and positive outcomes that can be achieved through a well-planned and executed finance transformation strategy.

Developing a winning finance transformation strategy is vital for organisations looking to navigate the future successfully. By aligning financial processes with strategic objectives, leveraging technology and automation, and embracing data analytics, organisations can optimise their operations, enhance decision-making capabilities, and drive sustainable growth. Assessing current finance processes, setting clear goals and objectives, developing a roadmap, implementing the strategy, and measuring success are key steps in this transformative journey. Real-life case studies provide valuable insights into the practical implementation and impact of finance transformation strategies. By adopting a holistic and strategic approach, organisations can develop a winning finance transformation strategy that positions them for success in the ever-evolving business landscape.

CHAPTER 4

Finance Leadership

FINANCE LEADERSHIP PLAYS A crucial role in the success of any organisation. As a finance leader, you are responsible for managing the financial health of the company. You oversee budgeting, financial planning, and analysis and make strategic decisions based on financial data. Your expertise is essential in guiding the organisation towards profitability and sustainable growth.

To excel in this role, you need to possess a diverse set of skills and qualities that go beyond number crunching. The modern finance leader needs to be a strategic thinker, an effective communicator, and a strong leader.

In my experience, the key qualities of a successful finance leader include:

Strategic Mindset: A successful finance leader has a clear understanding of the organisation's goals and aligns financial strategies accordingly. They analyse market trends, identify growth opportunities, and develop financial plans that support the company's long-term vision.

Analytical Skills: Finance leaders must have strong analytical skills to interpret complex financial data and make informed decisions. They should be able to identify trends, evaluate risks, and provide insights that drive financial performance.

Leadership Abilities: A finance leader is not only responsible for managing finances but also for leading a team. Strong leadership skills are crucial to motivate and inspire the finance team to achieve its objectives. Effective delegation, problem-solving, and decision-making are key qualities of a successful finance leader.

While finance leadership can be rewarding, it also comes with its fair share of challenges. Let's explore some common challenges faced by finance leaders and how to overcome them.

Changing Regulatory Landscape: Finance leaders need to stay updated with constantly evolving regulations and compliance requirements. It is essential to establish robust internal controls and maintain a strong ethical framework to navigate through regulatory complexities.

DOI: 10.1201/9781003514503-4

Managing Risk: Financial decisions involve inherent risks. Finance leaders need to develop risk management strategies and ensure that the organisation is adequately protected against potential financial threats. Regular risk assessments and proactive mitigation measures are essential to manage risk effectively.

Adapting to Technological Advancements: The finance industry is witnessing a rapid technological transformation. Finance leaders need to embrace new technologies such as AI, data analytics, and automation to streamline processes, improve efficiency, and generate accurate financial insights.

To become a masterful finance leader, you need to adopt key strategies that enhance your effectiveness and drive success:

Building a High-Performing Finance Team: A finance leader is only as good as their team. Building a high-performing finance team is essential to achieve financial goals and drive organisational success. Start by hiring talented professionals with diverse skill sets and a passion for finance. Encourage collaboration and foster a culture of continuous learning and development within the team. Provide opportunities for career growth and recognise and reward achievements. By building a strong and motivated team, you will be able to achieve exceptional results.

Developing Strong Communication and Interpersonal Skills: As a finance leader, effective communication is key in conveying financial information to stakeholders across the organisation. Develop strong communication and interpersonal skills to explain complex financial concepts in a clear and concise manner. Actively listen to others' perspectives and foster open and transparent communication channels. By building strong relationships with colleagues, you will gain their trust and support, enabling you to influence financial decisions effectively.

Leveraging Technology for Finance Leadership: Technology has revolutionised the finance industry, and finance leaders need to leverage it to their advantage. Embrace financial management software, data analytics tools, and automation to streamline processes, improve accuracy, and gain real-time insights. By harnessing the power of technology, you can make data-driven decisions, optimise financial performance, and drive innovation within your organisation.

Continuous Learning and Professional Development: In the rapidly evolving finance landscape, continuous learning and professional development are essential for finance leaders. Stay updated with industry trends, attend conferences and workshops, and pursue certifications to enhance your knowledge and skills. Encourage your team to do the same and promote a culture of lifelong learning within the organisation. By staying ahead of the curve, you can adapt to market changes and make strategic financial decisions.

The Importance of Ethics and Integrity: Ethics and integrity are fundamental pillars of finance leadership. As a finance leader, you have access to sensitive financial information and are responsible for maintaining the highest ethical standards. Upholding integrity ensures transparency, builds trust with stakeholders, and safeguards the company's reputation. Lead by example and establish a strong ethical framework within your team to foster a culture of accountability and integrity.

Mastering the art of finance leadership is a continuous journey that requires dedication, resilience, and a commitment to personal growth. Embrace challenges as opportunities for learning and growth, and continuously seek ways to enhance your skills and knowledge. Surround yourself with mentors and seek feedback to improve your leadership abilities. By combining technical expertise, strategic thinking, and strong interpersonal skills, you can become a masterful finance leader who drives financial success and leads their organisation towards a prosperous future.

Finance leadership is a multifaceted role that requires a diverse range of skills and qualities. By developing a strategic mindset, honing analytical skills, and cultivating strong leadership abilities, you can excel in this role. Overcome challenges by staying updated with regulations, managing risks effectively, and embracing technological advancements. Implement key strategies such as building a high-performing finance team, developing strong communication skills, leveraging technology, and investing in continuous learning and professional development. Uphold ethics and integrity to build trust and maintain a strong ethical framework within your organisation. With dedication and a commitment to personal growth, you can become a masterful finance leader who drives success and leads your organisation towards a prosperous future.

CHAPTER 5

Finance Operating Model

IN TODAY'S RAPIDLY EVOLVING business landscape, organisations need to have a strong and efficient finance operating model in place to navigate the complexities of financial management. The finance operating model serves as a blueprint for how financial operations are structured and executed within an organisation. It encompasses the processes, systems, roles, and responsibilities that enable effective financial management.

Harsono et al. (2024) argue fintech companies have strength in delivering impact in improving financial services, solving existing problems, delivering a more efficient and effective finance operating model, and providing paradigm shifts in the financial industry. They contend they have strength in optimising their finance operating model through the benefits of financial technology and, thereon, improving efficiency, accessibility, and innovation (Harsono and Suprapti, 2024).

In my experience, a well-designed finance operating model is crucial for streamlining financial operations, increasing efficiency, and driving better decision-making. It provides clarity and transparency, enabling organisations to optimise their financial resources and effectively manage risks. By understanding the finance operating model, organisations can identify areas for improvement and develop strategies to enhance their financial performance.

Streamlining financial operations is a critical priority for organisations seeking to achieve operational excellence and sustainable growth. Efficient financial operations enable organisations to make informed decisions, allocate resources effectively, and respond quickly to changing market dynamics. By streamlining financial operations, organisations can reduce costs, improve productivity, and enhance overall performance.

One of the key benefits of streamlining financial operations is the ability to gain real-time visibility into financial data. This enables organisations to generate accurate and timely financial reports, which are essential for effective decision-making. By eliminating manual and time-consuming processes, organisations can automate routine tasks, minimise errors, and free up resources for more strategic activities.

Furthermore, streamlining financial operations allows organisations to establish robust controls and compliance measures. This helps mitigate risks, ensure regulatory compliance,

DOI: 10.1201/9781003514503-5

and safeguard the organisation's financial assets. By implementing streamlined processes and leveraging technology solutions, organisations can enhance data accuracy, reduce fraud, and improve overall financial governance.

Developing a robust finance operating model can be a complex and challenging endeavour. Organisations often face several common challenges when embarking on this journey. In my experience, these challenges include:

Lack of Alignment: Different departments within the organisation may have different priorities and objectives, making it challenging to align financial operations across the organisation. This lack of alignment can result in inefficiencies, duplication of efforts, and inconsistent financial reporting.

Legacy Systems and Processes: Outdated systems and manual processes can hinder the development of an efficient finance operating model. These legacy systems may lack integration capabilities, making it difficult to streamline processes and obtain real-time financial insights.

Resistance to Change: Implementing a new finance operating model often requires changes in roles, responsibilities, and processes. Resistance to change from employees can impede the successful implementation of the new model and hinder the organisation's ability to achieve its financial goals.

Limited Resources: Developing a robust finance operating model requires dedicated resources, including skilled professionals, technology solutions, and financial investments. Limited resources can pose a significant challenge for organisations, particularly for smaller businesses with constrained budgets.

Despite these challenges, organisations can overcome them by following a structured and systematic approach to developing a finance operating model.

In my experience, in order to develop a *robust finance operating model*, organisations should follow a step-by-step approach that encompasses key stages. These stages include assessing the current state of financial operations, defining the desired future state, identifying gaps and areas for improvement, designing the new finance operating model, implementing and testing the new model, and monitoring and continuously improving it.

The first step in developing a robust finance operating model is to *assess the current state of financial operations*. This involves conducting a comprehensive review of existing processes, systems, roles, and responsibilities. Organisations should analyse key performance indicators, gather feedback from stakeholders, and identify pain points and areas of inefficiency.

During this assessment phase, organisations should also evaluate the effectiveness of existing controls, compliance measures, and governance frameworks. This will help identify any gaps or weaknesses that need to be addressed in the new finance operating model.

Once the current state of financial operations has been assessed, organisations can *define the desired future state*. This involves setting clear objectives, goals, and expectations for the new finance operating model. Organisations should consider industry best practices, benchmark against peer organisations, and align the future state with the overall strategic objectives of the organisation.

Defining the desired future state also involves identifying key performance indicators and metrics that will be used to measure the success of the new finance operating model. These metrics should be aligned with the organisation's financial goals and provide meaningful insights into the performance of financial operations.

After defining the desired future state, organisations should *identify gaps and areas for improvement* in their current finance operating model. This involves conducting a gap analysis to determine the key areas that need to be addressed in the new model. Organisations should consider factors such as process inefficiencies, a lack of automation, inadequate controls, and outdated technology.

By identifying these gaps, organisations can develop targeted strategies and initiatives to address them in the new finance operating model. This may involve redefining roles and responsibilities, implementing new systems and technologies, and enhancing controls and compliance measures.

Once the gaps and areas for improvement have been identified, organisations can *design the new finance operating model.* This involves developing a detailed plan that outlines the processes, systems, roles, and responsibilities of the new model. Organisations should consider factors such as process standardisation, automation, resource allocation, and technology integration.

During the design phase, organisations should also consider change management strategies to ensure a smooth transition to the new finance operating model. This may involve communicating the changes to employees, providing training and support, and addressing any concerns or resistance to change.

After designing the new finance operating model, organisations can proceed with its *implementation.* This involves executing the plan, deploying new systems and technologies, and assigning roles and responsibilities to employees. Organisations should closely monitor the implementation process, address any issues or challenges that arise, and ensure that the new model is effectively integrated into the organisation's operations.

Once the new finance operating model has been implemented, organisations should conduct thorough *testing* to validate its effectiveness. This may involve conducting pilot tests, gathering feedback from stakeholders, and making necessary adjustments and refinements.

The final step in developing a robust finance operating model is to establish a *monitoring and continuous improvement framework.* Organisations should regularly monitor the performance of the new model, track key performance indicators, and gather feedback from stakeholders. This will help identify any areas of improvement and enable organisations to make data-driven decisions to enhance the effectiveness of financial operations.

Organisations should also foster a culture of continuous improvement by encouraging feedback, promoting innovation, and implementing best practices. This will ensure that the finance operating model remains agile and adaptable to changing business requirements.

Developing a robust finance operating model requires careful planning, execution, and ongoing management. To ensure successful finance transformation, organisations should consider the following key factors:

Executive Sponsorship: Finance transformation initiatives require strong executive sponsorship and support. Senior leaders should champion the transformation efforts, provide the necessary resources, and communicate the strategic importance of the finance operating model to the organisation.

Collaboration: Developing a finance operating model is a cross-functional effort that requires collaboration and coordination among various departments and stakeholders. Organisations should foster a collaborative culture, establish clear communication channels, and involve key stakeholders throughout the transformation journey.

Change Management: Implementing a new finance operating model often involves significant changes in roles, responsibilities, and processes. Organisations should develop a robust change management plan that includes communication, training, and support to ensure a smooth transition and minimise resistance to change.

Technology Enablement: Leveraging technology solutions is essential for streamlining financial operations and enabling effective decision-making. Organisations should invest in modern finance systems, automation tools, and analytics capabilities to enhance the efficiency and effectiveness of financial operations.

Continuous Improvement: Developing a finance operating model is an ongoing process that requires continuous improvement. Organisations should establish a feedback loop, monitor key performance indicators, and regularly evaluate the effectiveness of the finance operating model. This will enable organisations to identify areas for improvement and make data-driven decisions to optimise financial operations.

To illustrate the benefits and outcomes of developing a robust finance operating model, let's explore two case studies of successful finance operating model transformations.

Samsung: Samsung, a multinational manufacturing company, embarked on a finance transformation journey to streamline its financial operations and enhance its decision-making capabilities. The organisation conducted a comprehensive assessment of its current state of financial operations, which revealed several inefficiencies and gaps in processes and systems.

Based on the assessment findings, Samsung defined a desired future state that focused on process standardisation, automation, and enhanced controls. The organisation redesigned its finance operating model, implementing new systems and technologies to automate routine tasks and improve data accuracy.

Through the implementation and testing phases, Samsung achieved significant improvements in financial reporting accuracy, reduced manual effort, and enhanced compliance measures. The organisation also established a robust monitoring and continuous improvement framework, which enabled ongoing enhancements to the finance operating model.

As a result of the finance operating model transformation, Samsung experienced improved decision-making capabilities, reduced costs, and increased productivity. The organisation was able to allocate resources more effectively, respond quickly to market changes, and achieve sustainable growth.

Lloyds Bank: Lloyds Bank, a financial services organisation, recognised the need to develop a robust finance operating model to address the challenges of a rapidly changing regulatory landscape. The organisation conducted a thorough assessment of its current financial operations, which revealed potential improvements in compliance measures and outdated systems.

Based on the assessment findings, Lloyds Bank defined a desired future state that focused on enhancing compliance, improving data accuracy, and increasing operational efficiency. The organisation redesigned its finance operating model, implementing new technologies to automate compliance processes and enhance data governance.

During the implementation and testing phase, Lloyds Bank experienced improved compliance measures, enhanced data accuracy, and reduced regulatory risks. The organisation also established a monitoring and continuous improvement framework, which enabled ongoing enhancements to the finance operating model to adapt to evolving regulatory requirements.

As a result of the finance operating model transformation, Lloyds Bank achieved improved regulatory compliance, reduced operational risks, and enhanced customer trust. The organisation was able to streamline its financial operations, allocate resources more effectively, and achieve sustainable growth in a highly regulated industry.

In my experience, developing a robust finance operating model requires access to key resources and tools. Here are a few resources that can help organisations in their finance transformation journey:

Finance Transformation Framework: A comprehensive framework that provides guidance on developing a finance operating model, including templates, best practices, and case studies.

Finance Systems and Technology Solutions: Modern finance systems and technology solutions that enable organisations to automate routine tasks, enhance data accuracy, and improve decision-making capabilities.

Financial Performance Metrics: Key performance indicators and metrics that organisations can use to measure the effectiveness of their finance operating model and track financial performance.

Training and Development Programs: Training programs and certifications that help finance professionals develop the skills and knowledge required for a successful finance operating model transformation.

Industry Associations and Networks: Professional associations and networks that provide access to industry experts, thought leadership, and peer-to-peer learning opportunities.

By leveraging these resources and tools, organisations can enhance their finance operating model and achieve their financial goals.

Robustness is vital. Developing a robust finance operating model is essential for organisations seeking to streamline financial operations, increase efficiency, and drive better decision-making. By following a step-by-step approach and considering key considerations, organisations can successfully develop and implement a finance operating model that aligns with their strategic objectives.

Through continuous monitoring and improvement, organisations can optimise their financial operations, reduce costs, improve compliance measures, and achieve sustainable growth. By leveraging the right resources and tools, organisations can enhance their finance operating model and position themselves for long-term success in today's dynamic business environment.

5.1 STREAMLINING THE ORDER-TO-CASH PROCESS

As a finance professional, I understand the criticality of having an efficient order-to-cash process. The order-to-cash process is the backbone of any organisation's financial operations, as it encompasses all the activities involved in receiving and fulfilling customer orders, invoicing, and collecting payments. Streamlining this process is essential for businesses to operate smoothly and maximise their financial performance.

One of the primary reasons why streamlining the order-to-cash process is crucial is because it directly impacts cash flow. By ensuring that orders are processed quickly and accurately, invoices are generated promptly, and payments are collected on time, businesses can improve their cash flow position. This, in turn, provides them with the necessary liquidity to meet their financial obligations, invest in growth opportunities, and respond to unforeseen challenges.

Moreover, a streamlined order-to-cash process enhances customer satisfaction. By reducing the time taken to process orders and deliver products or services, businesses can meet customer expectations and build strong relationships. It also minimises errors and discrepancies in invoices, leading to fewer disputes and improved customer trust. Ultimately, satisfied customers are more likely to become repeat customers and recommend the business to others, driving revenue growth.

Efficiency in the order-to-cash process also enables better decision-making. By having accurate and timely financial information, businesses can analyse their performance, identify trends, and make informed strategic decisions. This includes determining pricing strategies, allocating resources effectively, and identifying areas for improvement. A streamlined order-to-cash process provides finance professionals with reliable data, allowing them to contribute to the organisation's overall success.

To effectively streamline the order-to-cash process, it is important to understand its key components. The order-to-cash process typically consists of the following stages:

Order Entry and Processing: This stage involves receiving and entering customer orders into the system. It includes verifying order details, checking product availability, and scheduling delivery or service provision.

Inventory Management: Once orders are confirmed, businesses need to manage their inventory effectively to fulfil customer demands. This includes maintaining optimal stock levels, tracking inventory movements, and coordinating with suppliers if additional stock is required.

Invoicing and Billing: Invoicing is a critical step in the order-to-cash process. It involves generating accurate and timely invoices based on the order details. The invoices should include all relevant information, such as product or service descriptions, quantities, prices, and any applicable taxes or discounts.

Payment Processing and Collection: After invoices are sent to customers, businesses need to ensure timely payment collection. This involves offering various payment options, tracking payment receipts, following up on outstanding invoices, and managing any disputes or discrepancies.

Accounts Receivable Management: Efficient accounts receivable management is essential for maintaining a healthy cash flow. It includes monitoring customer payment behaviours, establishing credit policies, and implementing effective collection strategies to minimise bad debts.

By understanding these key components, businesses can identify potential bottlenecks and inefficiencies in their order-to-cash process. This insight is crucial for developing strategies to streamline operations and enhance overall efficiency.

Despite the importance of streamlining the order-to-cash process, there are several challenges that businesses often face. These challenges can hinder efficiency and impact the organisation's financial performance. Some of the common challenges include:

Manual Processes and Lack of Automation: Many businesses still rely on manual processes, such as manual data entry or paper-based invoices. These processes are time-consuming, error-prone, and limit scalability. Implementing automation technologies, such as electronic data interchange (EDI) or robotic process automation (RPA), can significantly improve efficiency and reduce errors.

Lack of Integration between Systems: Inefficient integration between different systems, such as order management, inventory management, and accounting software, can lead to data discrepancies, delays, and errors. Implementing integrated enterprise resource planning (ERP) systems can ensure seamless data flow and improve overall process efficiency.

Complex Pricing and Discount Structures: Businesses that offer complex pricing structures or discounts often face challenges in accurately calculating invoices and managing pricing changes. Simplifying pricing structures and automating the calculation of discounts can streamline the invoicing process and reduce errors.

Ineffective Communication and Collaboration: Lack of communication and collaboration between different departments, such as sales, finance, and customer service, can lead to delays, errors, and customer dissatisfaction. Implementing cross-functional collaboration tools and fostering a culture of communication can improve overall process efficiency.

Addressing these challenges is crucial for unlocking efficiency in the order-to-cash process. By identifying and implementing appropriate solutions, businesses can streamline operations, reduce costs, and improve customer satisfaction.

To unlock efficiency in the order-to-cash process, businesses can adopt several strategies:

Implementing a Streamlined Procure to Pay Process: The procure-to-pay process is closely linked to the order-to-cash process and plays a vital role in overall efficiency. By streamlining the procure-to-pay process, businesses can ensure the timely procurement of goods or services, accurate recording of expenses, and seamless integration with the order-to-cash process.

One way to streamline the procure-to-pay process is by implementing e-procurement systems. These systems automate the procurement process, from requisition to payment, reducing manual errors and processing time. They also provide visibility into procurement activities, enabling better financial planning and cost control.

Another strategy is to establish strong *vendor management practices*. This includes conducting thorough vendor assessments, negotiating favourable terms and conditions, and establishing clear communication channels. By building strong relationships with vendors, businesses can optimise procurement processes, negotiate better pricing, and reduce supply chain risks.

Optimising Fixed Assets Management: Fixed assets, such as buildings, machinery, or vehicles, are significant investments for businesses. Efficient management of these assets is essential for maximising their utilisation and minimising costs. By optimising fixed asset management, businesses can improve their financial performance and streamline the order-to-cash process.

One approach is to implement a robust asset-tracking system. This system should accurately record asset details, including acquisition costs, depreciation, maintenance schedules, and disposal information. By having real-time visibility into asset data, businesses can make informed decisions regarding asset utilisation, maintenance, and replacement.

Additionally, businesses can leverage technology to improve fixed asset management. Internet of Things devices can be used to monitor asset performance, detect maintenance issues proactively, and optimise asset usage. This minimises downtime, reduces maintenance costs, and improves overall operational efficiency.

Streamlining the Close, Consolidate, and Report Process: The close, consolidate, and report process is a critical stage in the order-to-cash process. It involves consolidating financial data, preparing financial statements, and reporting to stakeholders. Streamlining this process is essential for accurate and timely financial reporting, compliance with regulatory requirements, and effective decision-making.

One strategy for streamlining the close, consolidate, and report process is to implement financial consolidation software. This software automates the consolidation process, eliminates manual errors, and provides real-time visibility into financial performance. It also enables businesses to generate accurate financial statements, such as balance sheets and income statements, quickly and efficiently.

Another approach is to establish standardised reporting templates and processes. By defining clear reporting guidelines and templates, businesses can ensure consistency and reduce the time taken to prepare reports. Implementing reporting tools or dashboards can further enhance efficiency by providing stakeholders with real-time access to relevant financial information.

To improve efficiency in finance operations, businesses can leverage various tools and technologies:

ERP Systems: ERP systems integrate different finance and accounting functions into a single platform, streamlining processes and improving data accuracy. These systems automate tasks such as invoice processing, payment collection, and financial reporting, reducing manual effort and errors.

RPA: RPA technology can automate repetitive and rule-based finance tasks, such as data entry or reconciliation. By implementing RPA, businesses can free up finance professionals' time, reduce errors, and improve overall process efficiency.

EDI: EDI enables the electronic exchange of data between different systems, such as order management and invoicing systems. Implementing EDI eliminates manual data entry, reduces errors, and improves data accuracy, leading to streamlined operations.

Cloud-Based Financial Management Systems: Cloud-based financial management systems provide businesses with real-time access to financial data and tools from any location. These systems offer scalability, data security, and collaboration capabilities, enabling finance professionals to work efficiently and effectively.

Streamlining the order-to-cash process is essential for businesses to unlock efficiency in their finance operations. By optimising key components of the order-to-cash process, such as procure-to-pay, fixed asset management, and the close, consolidate, and report processes, businesses can improve cash flow, enhance customer satisfaction, and enable better decision-making.

Implementing tools and technologies, such as ERP systems, RPA, EDI, and cloud-based financial management systems, further improves efficiency and reduces manual effort. These tools automate tasks, improve data accuracy, and provide real-time visibility into financial performance.

A streamlined finance operating model not only enables businesses to operate more efficiently but also positions them for growth and success. By investing in streamlining the order-to-cash process and leveraging appropriate tools and technologies, businesses can maximise their financial performance, reduce costs, and stay ahead of the competition.

5.2 SHARED SERVICES

In today's rapidly changing business landscape, organisations are constantly seeking ways to optimise their operations and improve efficiency. One approach that has gained significant traction is the implementation of shared services. Shared services involve consolidating and centralising certain functions, such as finance, human resources, or IT, to create a more streamlined and cost-effective operating model.

By centralising these functions, organisations can eliminate redundancies, reduce costs, and improve overall service delivery. Shared services can be implemented within an organisation, serving multiple business units, or they can be outsourced to a third-party service provider.

Regardless of the approach, successful implementation relies on a well-designed finance operating model.

Implementing a *shared services model* offers numerous benefits to organisations. Firstly, it allows for standardisation and consistency in processes and procedures. By centralising finance functions, organisations can establish standardised workflows, policies, and controls, ensuring greater accuracy and compliance.

Secondly, shared services drive economies of scale. By consolidating resources and leveraging technology, organisations can achieve cost savings through reduced overhead, improved purchasing power, and optimised resource allocation. This, in turn, frees up resources that can be allocated to strategic initiatives or revenue-generating activities.

Finally, implementing shared services enhances service quality. By centralising expertise and resources, organisations can provide faster response times, improved data accuracy, and better customer service. This enables business units to focus on their core competencies while relying on the shared services team for critical support.

A crucial element in designing successful shared services is understanding the *finance operating model*. The finance operating model defines how finance functions are structured, the roles and responsibilities of various stakeholders, and the processes and systems that support the delivery of finance services.

There are typically three main components of a finance operating model: the governance structure, the process framework, and the technology infrastructure. The governance structure outlines decision-making authority, reporting lines, and performance metrics. The process framework defines the end-to-end processes, including transactional activities, financial reporting, and compliance. The technology infrastructure involves the systems, tools, and data management processes that support finance operations.

Designing a successful shared services model requires careful consideration of several key factors. Firstly, organisations need to define the scope and scale of the shared services. This involves determining which finance functions are suitable for centralisation and assessing the volume and complexity of the transactions involved.

Secondly, organisations must establish clear governance and accountability structures. This includes defining the roles and responsibilities of the shared services team, as well as the relationship between the shared services team and the business units they serve. Effective governance ensures alignment, transparency, and accountability throughout the shared services model.

Thirdly, organisations need to prioritise change management and stakeholder engagement. Implementing a shared services model often involves significant changes to processes, systems, and roles. It is crucial to proactively communicate the benefits of the shared services model, address concerns, and provide adequate training and support to affected employees.

To streamline the finance operating model for shared services, organisations should focus on several areas. Firstly, process standardisation is essential. By establishing consistent and efficient processes, organisations can eliminate redundancies, reduce errors, and improve overall efficiency. This involves documenting and streamlining workflows, implementing standardised policies and controls, and leveraging technology to automate manual tasks.

Secondly, organisations should invest in technology infrastructure that supports the shared services model. This includes implementing an integrated financial management system, such as an ERP system, which enables seamless data sharing, real-time reporting, and process automation. Additionally, organisations should leverage advanced analytics and business intelligence tools to gain insights into financial performance and identify areas for improvement.

Finally, organisations should prioritise continuous improvement and innovation. By regularly reviewing and optimising processes, organisations can identify opportunities to streamline operations, increase efficiency, and enhance service delivery. This involves leveraging data and analytics to identify bottlenecks and inefficiencies, implementing best practices from industry benchmarks, and fostering a culture of innovation and continuous learning within the shared services team.

Implementing a shared services model requires careful planning and execution. Firstly, organisations need to establish a clear implementation roadmap. This involves defining the project scope, setting objectives and milestones, and allocating resources and budgets.

Additionally, organisations should establish a dedicated project team with the necessary skills and expertise to drive the implementation process.

Secondly, organisations should conduct a thorough assessment of their existing finance operations. This involves analysing current processes, systems, and capabilities, identifying areas for improvement, and developing a detailed transition plan. The transition plan should outline the sequence of activities, the timeline, and the resources required for the implementation.

Thirdly, organisations should proactively manage change and stakeholder engagement throughout the implementation process. This includes communicating the benefits of the shared services model, addressing concerns and resistance, and providing adequate training and support to affected employees. Change management should be an ongoing process, with regular updates and feedback loops to ensure a smooth transition.

Outsourcing the design and implementation of shared services can offer additional benefits, such as accessing specialised expertise, accelerating implementation timelines, and reducing upfront investment. However, it also presents unique challenges and risks that need to be managed effectively.

To ensure successful outsourcing, organisations should follow several best practices. Firstly, organisations should conduct thorough due diligence when selecting a service provider. This includes assessing the provider's track record, capabilities, and financial stability, as well as their cultural fit with the organisation.

Secondly, organisations should establish clear service level agreements and performance metrics with the service provider. This ensures that expectations are clearly defined and monitored and that the service provider is held accountable for delivering the agreed-upon outcomes.

Finally, organisations should establish a governance framework to oversee the outsourcing relationship. This involves defining the roles and responsibilities of both the organisation and the service provider, establishing regular communication channels, and conducting periodic performance reviews. Effective governance ensures alignment, transparency, and continuous improvement in the outsourcing relationship.

While implementing shared services can offer significant benefits, organisations must also be aware of the challenges and risks involved. One common challenge is resistance to change.

Implementing a shared services model often involves significant changes to processes, systems, and roles, which can create uncertainty and resistance among employees. Effective change management and stakeholder engagement are crucial to address these challenges.

Another challenge is ensuring data security and privacy. Centralising finance functions requires sharing sensitive financial data across multiple business units or with a third-party service provider. Organisations must implement robust data security measures, including encryption, access controls, and regular audits, to ensure the confidentiality and integrity of financial information.

Additionally, organisations must carefully manage the transition from decentralised to centralised finance operations. This involves ensuring continuity of service during the transition, providing adequate training and support to affected employees, and addressing any potential disruptions or conflicts that may arise.

Several organisations have successfully implemented shared services models, achieving significant cost savings, improved service quality, and enhanced efficiency. One example is a global manufacturing company that centralised its finance functions, including accounts payable, accounts receivable, and financial reporting. By streamlining processes, implementing a robust technology infrastructure, and leveraging outsourcing partners, the company achieved cost savings of over 30% and improved data accuracy and timeliness.

Another example is a multinational technology company that implemented a shared services model for its human resources functions. By centralising HR processes, including recruitment, payroll, and benefits administration, the company achieved greater consistency in service delivery, reduced administrative burden on business units, and improved compliance with regulatory requirements.

These case studies highlight the importance of a well-designed finance operating model, effective change management, and continuous improvement in achieving successful shared service implementation.

Implementing a shared services model can offer significant benefits to organisations, including cost savings, improved service quality, and enhanced efficiency. However, successful implementation relies on a well-designed finance operating model that streamlines processes, leverages technology, and prioritises continuous improvement.

By carefully considering key factors such as scope and scale, governance and accountability, and stakeholder engagement, organisations can design a shared services model that meets their specific needs and objectives. Additionally, by investing in process standardisation, technology infrastructure, and continuous improvement, organisations can achieve even greater efficiency and effectiveness in their finance operations.

Whether implementing shared services internally or outsourcing to a service provider, organisations must proactively manage change, address challenges and risks, and foster a culture of innovation and continuous learning. By doing so, organisations can achieve success in their shared service design and implementation, driving greater operational efficiency and enabling a focus on strategic value-add activities.

CHAPTER **6**

Refine Process and Policy

FINANCE IS THE BACKBONE of any organisation, whether it be a multinational corporation or a small start-up. It encompasses the management of money, investments, and
other financial resources. Understanding finance is crucial for individuals and businesses
alike, as it enables us to make informed decisions about our financial well-being. From
budgeting to investing, finance plays a pivotal role in our lives.

Financial success is not solely dependent on luck; it is a result of careful planning,
strategic decision-making, and adherence to effective processes and policies. These processes and policies provide a framework that guides the financial management of an
organisation, ensuring its sustainability and growth. By demystifying the importance of
the process and policy in finance, we can gain a deeper understanding of how they contribute to financial success.

In the world of finance, process and policy are two key pillars that support an organisation's financial success. A process refers to a series of steps or actions that are undertaken
to achieve a specific financial objective. It provides a structured approach to managing
finances, ensuring consistency and efficiency. On the other hand, a policy sets the guidelines and boundaries within which financial decisions are made. It establishes the rules
and principles that govern financial activities, ensuring transparency and accountability.

The process and policy work hand in hand to create a robust financial management system. The process outlines the steps to be followed, while the policy sets the rules to be adhered
to during these steps. Together, they provide a framework that ensures financial activities
are carried out in a controlled and systematic manner. By establishing clear processes and
policies, organisations can mitigate risks, prevent fraud, and promote financial stability.

Financial controls are an integral part of the process and policy framework. They are
mechanisms put in place to monitor and regulate financial activities, ensuring compliance
with established policies and procedures. Financial controls encompass a wide range of activities, including budgeting, cash flow management, financial reporting, and internal audits.

The primary objective of financial controls is to safeguard the organisation's assets and
ensure the accuracy and reliability of financial information. They act as checks and bal-
ances, preventing errors, fraud, and mismanagement. In my experience, by implementing

DOI: 10.1201/9781003514503-6

effective financial controls, organisations can minimise financial risks, make informed decisions, and maintain the trust of stakeholders.

Effective financial processes and policies consist of several key elements. These elements provide the foundation for a robust financial management system. They include:

Clear Objectives: Financial processes and policies should align with the overall goals and objectives of the organisation. They should be designed to support strategic decision-making and drive financial success.

Defined Roles and Responsibilities: Each financial process and policy should clearly define the roles and responsibilities of individuals involved. This ensures accountability and prevents confusion or duplication of efforts.

Documentation and Record-Keeping: Proper documentation and record-keeping are essential to maintain transparency and accountability. Financial processes and policies should outline the requirements for documentation and establish procedures for record-keeping.

Regular Monitoring and Review: Financial processes and policies should be regularly monitored and reviewed to ensure their effectiveness. This includes conducting internal audits, analysing financial data, and identifying areas for improvement.

Implementing effective financial controls offers numerous benefits for organisations. These benefits extend beyond financial stability and include:

Risk Mitigation: Financial controls help identify and mitigate risks, minimising the likelihood of financial loss or fraud. By implementing controls such as segregation of duties and regular audits, organisations can safeguard their assets.

Decision-Making Support: Financial controls provide accurate and reliable financial information, enabling informed decision-making. By having a clear understanding of the organisation's financial position, management can make strategic choices that drive growth and profitability.

Compliance with Regulations: Effective financial controls ensure compliance with applicable laws and regulations. This helps organisations avoid legal penalties and maintain their reputation.

Improved Efficiency: Well-designed financial controls streamline processes, reducing redundancies and inefficiencies. This leads to cost savings and improves operational efficiency.

Several successful companies serve as examples of the benefits of implementing strong financial controls. One such company is Google. Google has a rigorous financial control system that ensures accuracy and transparency in its financial reporting. The company's financial controls include regular internal audits, segregation of duties, and strict adherence to accounting standards. These controls have contributed to Google's financial success and its position as one of the most valuable companies in the world.

Another example is Procter & Gamble (P&G). P&G is known for its robust financial controls, which have helped the company maintain financial stability and drive growth. P&G's financial controls include detailed financial planning and budgeting, regular performance tracking, and stringent risk management practices. These controls have enabled P&G to make informed decisions, optimise resource allocation, and achieve sustainable financial success.

Implementing financial processes and policies is not without its challenges. Organisations often face several common pitfalls during the implementation phase. In my experience, some of these challenges include:

Resistance to Change: Implementing new financial processes and policies may face resistance from employees who are accustomed to existing practices. Overcoming resistance requires effective change management and clear communication about the benefits of the changes.

Lack of Resources: Designing and implementing effective financial processes and policies require dedicated resources, including time, expertise, and technology. Organisations should ensure they have the necessary resources to support the implementation process.

Inadequate Training and Education: Employees need to be trained on the new financial processes and policies to ensure their successful implementation. Inadequate training can lead to misunderstandings, errors, and non-compliance.

Lack of Continuous Improvement: Financial processes and policies should be regularly reviewed and improved to adapt to changing business environments. Failure to continuously improve can lead to outdated practices and inefficiencies.

To overcome these challenges and ensure the successful implementation of financial processes and policies, organisations should follow best practices. Some of these best practices include:

Engage Stakeholders: Involve key stakeholders in the design and implementation process to ensure their buy-in and support. This includes employees, management, and external consultants or auditors.

Communicate Effectively: Clearly communicate the purpose, benefits, and expectations of the new financial processes and policies to all stakeholders. This helps create awareness and understanding, reducing resistance to change.

Provide Comprehensive Training: Develop and deliver comprehensive training programmes to educate employees on the new financial processes and policies. This includes providing hands-on training, reference materials, and ongoing support.

Monitor and Evaluate: Regularly monitor and evaluate the effectiveness of financial controls to identify areas for improvement. This includes conducting internal audits, analysing financial data, and soliciting feedback from stakeholders.

Evaluating the Effectiveness of financial controls is crucial to ensure their ongoing relevance and efficiency. Organisations can use various methods to evaluate their financial controls, including:

Internal Audits: Conducting regular internal audits helps identify any weaknesses or gaps in financial controls. Internal auditors can assess compliance with policies, review documentation, and make recommendations for improvement.

Key Performance Indicators (KPIs): Establishing KPIs related to financial controls allows organisations to track their performance and identify areas for improvement. KPIs can include metrics such as the accuracy of financial reporting, timeliness of financial transactions, and effectiveness of risk management practices.

Feedback from Stakeholders: Gather feedback from stakeholders, including employees, management, and external auditors or consultants. This feedback can provide valuable insights into the effectiveness of financial controls and identify areas for enhancement.

Benchmarking: Compare financial controls against industry best practices or peer organisations to identify areas for improvement. Benchmarking allows organisations to learn from others and adopt proven strategies.

Finance plays a crucial role in the success of organisations. The process and policy framework provides the structure and guidelines necessary to manage finances effectively. By implementing robust financial controls, organisations can mitigate risks, make informed decisions, and drive financial success.

However, it is important to recognise that financial processes and policies are not static. They should be regularly reviewed and improved to adapt to changing business environments and new challenges. Continuous improvement is key to maintaining the effectiveness of financial controls and ensuring their alignment with organisational goals.

By embracing best practices, monitoring performance, and seeking feedback from stakeholders, organisations can design and implement financial processes and policies that contribute to their long-term success. The demystification of finance and the understanding of the process and policy framework empower individuals and businesses to navigate the complex world of finance with confidence.

Financial transformation is a critical process for any organisation aiming to improve its financial performance and achieve long-term success. In today's dynamic business environment, organisations must adapt and evolve to stay competitive. CFOs must understand the importance of financial transformation, with a specific focus on the role of internal audit in maximising efficiency throughout the transformation journey.

Internal audit plays a crucial role in financial transformation by providing independent and objective assurance to the organisation's management and stakeholders. The internal audit function evaluates the effectiveness of internal controls, risk management processes, and governance structures. By assessing the existing financial processes and identifying areas for improvement, internal audit helps organisations streamline their operations and enhance their financial performance.

During financial transformation, internal audit serves as a trusted advisor, supporting the organisation's management in making informed decisions. By conducting comprehensive risk assessments and audits, internal audit identifies potential risks and ensures that adequate controls are in place to mitigate those risks. Furthermore, internal audit provides valuable insights and recommendations to enhance the organisation's financial processes, enabling it to navigate the transformation journey effectively.

Incorporating internal audit in the financial transformation process offers numerous benefits for organisations. Firstly, internal audit brings an independent and objective perspective, ensuring that the transformation initiatives are aligned with the organisation's strategic goals. By providing unbiased assessments and recommendations, internal audit helps organisations avoid potential pitfalls and optimise their transformation efforts.

Secondly, internal audit enhances the overall governance and control environment during financial transformation. Through continuous monitoring and evaluation of internal controls, internal audit helps organisations reduce the risk of fraud, errors, and non-compliance. This proactive approach ensures that the transformation process is conducted in a secure and controlled manner, safeguarding the organisation's assets and reputation.

Lastly, an internal audit provides assurance to stakeholders, including the board of directors, investors, and regulatory bodies. By demonstrating the effectiveness of the organisation's financial processes and controls, internal audit instils confidence in the transformation initiatives. This assurance is vital for maintaining stakeholder trust and ensuring a smooth transition throughout the financial transformation journey.

The internal audit function assumes various responsibilities during financial transformation to ensure its success. Firstly, an internal audit conducts a comprehensive assessment of the organisation's existing financial processes, controls, and risk management frameworks. This assessment helps identify areas of improvement and sets the foundation for the transformation roadmap.

Secondly, internal audit collaborates with the organisation's management and other stakeholders to develop and implement robust internal controls and risk management frameworks. By aligning the transformation initiatives with industry best practices and regulatory requirements, internal audit helps organisations build a solid foundation for sustainable growth.

Furthermore, internal audit continuously monitors and evaluates the effectiveness of the financial transformation initiatives. This monitoring ensures that the implemented changes are delivering the expected results and enables timely adjustments if required. By providing regular feedback and reporting, internal audit keeps the organisation's management informed about the progress of the transformation journey.

In today's digital age, organisations can leverage technology to enhance the efficiency and effectiveness of their internal audit processes during financial transformation. Automated audit tools and software can streamline data collection, analysis, and reporting, reducing manual efforts and improving audit productivity.

By utilising data analytics and artificial intelligence, internal audit can gain deeper insights into the organisation's financial processes and controls. Advanced analytics tools can analyse large volumes of data quickly, identifying patterns, anomalies, and potential risks. This enables internal audit to provide more targeted recommendations for process improvements and risk mitigation.

Additionally, technology enables real-time monitoring and reporting, providing instant visibility into the organisation's financial performance and potential risks. Through dashboards and interactive reports, internal audit can track the progress of the transformation initiatives and identify any deviations or areas requiring immediate attention. This proactive approach ensures that the organisation stays on track and can address emerging issues promptly.

Delegation of authority is a critical aspect of financial transformation, enabling organisations to distribute decision-making responsibilities effectively. During transformation, management must delegate authority to different levels within the organisation, empowering employees to make informed decisions in line with the transformation objectives.

Delegation of authority allows for faster decision-making, as decisions can be made at the appropriate level without unnecessary delays. This promotes agility and responsiveness, enabling organisations to adapt to changing market conditions and capitalise on emerging opportunities.

However, effective delegation requires clear guidelines, accountability, and oversight. Internal audit plays a vital role in ensuring that the delegation process is well-defined, transparent, and aligned with the organisation's strategic goals. By conducting periodic reviews and audits, internal audit verifies that the delegated authorities are being exercised appropriately and that the necessary controls are in place to mitigate any associated risks.

To drive financial transformation successfully, organisations should adopt *best practices in internal audit and delegation*. Firstly, organisations should establish a robust internal audit charter that clearly defines the objectives, scope, and authority of the internal audit function. This charter should align with industry standards and regulatory requirements, ensuring that internal audit operates effectively and independently.

Secondly, organisations should implement a risk-based audit approach, focusing on areas with the highest potential impact on the transformation process. By prioritising audits based on risk, internal audit can allocate its resources efficiently and provide meaningful insights to management. This approach ensures that internal audit contributes to the organisation's strategic objectives and delivers value throughout the transformation journey.

Moreover, organisations should foster a culture of accountability and continuous improvement. Internal audit should collaborate with the organisation's management to develop KPIs and metrics that measure the effectiveness of the transformation initiatives. These metrics enable internal audit to track progress, identify areas for improvement, and provide timely recommendations to enhance the transformation process.

To illustrate the impact of internal audit on financial transformation, let's examine two case studies of organisations that successfully integrated internal audit in their transformation journeys.

Mercedes-Benz Group: Mercedes-Benz Group, a multinational manufacturing company, embarked on a financial transformation initiative to optimise its operational costs and improve profitability. Internal audit played a crucial role in this transformation journey by conducting a comprehensive assessment of the company's financial processes, identifying inefficiencies, and recommending process improvements.

By leveraging technology, internal audit automated several manual processes, such as data collection and analysis. This automation significantly improved audit efficiency and allowed internal audit to focus on value-added activities such as data interpretation and strategic recommendations.

Furthermore, internal audit collaborated closely with the organisation's management to develop a robust delegation framework. This framework clearly defined the decision-making authorities at various levels within the organisation, ensuring accountability and transparency. Internal audit conducted periodic audits to verify the effectiveness of the delegation process and provide feedback for continuous improvement.

As a result of the successful integration of internal audit in the financial transformation process, Mercedes-Benz Group achieved significant cost savings, improved operational efficiency, and enhanced profitability. The organisation's stakeholders, including investors and regulatory bodies, gained confidence in the transformation initiatives, leading to increased trust and support.

Citigroup: Citigroup, a financial services firm, underwent a digital transformation to enhance its customer experience and improve its competitiveness. Internal audit played a pivotal role in this transformation journey by assessing the organisation's existing digital infrastructure, identifying potential risks, and recommending cybersecurity measures.

Internal audit collaborated with the organisation's IT department to implement advanced analytics tools and real-time monitoring systems. These technological enhancements enabled internal audit to identify emerging risks promptly and provide timely recommendations to mitigate them. Additionally, internal audit conduct periodic audits to verify the effectiveness of the implemented cybersecurity measures and ensure compliance with industry standards.

Through the successful integration of internal audit in the financial transformation process, Citigroup achieved a seamless digital transition, enhanced customer satisfaction, and strengthened its competitive position. The organisation's management and stakeholders recognised the value of internal audit in maintaining a secure digital environment, fostering trust, and ensuring regulatory compliance.

Training and development opportunities must be provided by organisations to equip internal auditors with the necessary skills and knowledge for financial transformation. These opportunities enable internal auditors to stay updated with the latest industry trends, regulatory requirements, and technological advancements.

Organisations can offer training programmes and certifications focused on financial transformation, internal controls, risk management, and data analytics. These programmes enhance internal auditors' technical expertise and enable them to provide valuable insights and recommendations throughout the transformation journey.

Moreover, organisations should encourage internal auditors to participate in industry conferences, seminars, and workshops. These platforms provide opportunities for networking, knowledge sharing, and learning from industry experts. By staying connected with the broader internal audit community, internal auditors can gain valuable insights and best practices that can be applied in the context of financial transformation.

Internal audit and delegations can play a vital role in maximising efficiency during financial transformation. By providing independent and objective assurance, internal audit helps organisations streamline their financial processes, enhance their control environment, and achieve their transformation objectives. Through effective delegation of authority and the use of technology, this contributes to the success of the transformation process. By adopting best practices, organisations can integrate internal audit effectively and drive their financial transformation initiatives to achieve sustainable growth.

Finance Technology

I N TODAY'S RAPIDLY EVOLVING business landscape, finance transformation has become a critical aspect for organisations aiming to stay competitive. The advent of emerging finance technologies has revolutionised the way financial processes are carried out, enhancing efficiency, accuracy, and decision-making capabilities. These technologies encompass a wide range of tools and platforms that streamline financial operations, automate manual tasks, and provide real-time insights. From artificial intelligence and machine learning to blockchain and robotic process automation (RPA), finance technologies are shaping the future of finance departments across industries.

Pelykh (2020) contends that the rapid development of digital technologies is causing a paradigm shift in financial services and argues for the case for deep integration of digital technologies in the financial sector, which will lead to an increase in the quality and availability of services as well as the formation of new rules that ensure healthy competition between market participants (Pelykh, 2020).

There are key drivers that are fuelling the need for finance transformation through the adoption of emerging finance technologies. First and foremost, the increasing complexity of financial transactions and the ever-growing volumes of data have made it imperative for organisations to leverage technology to analyse and interpret this information accurately and efficiently. Finance technologies enable organisations to access real-time data, perform advanced analytics, and generate actionable insights, thereby empowering finance professionals to make informed decisions.

Another significant driver of finance transformation is the need for enhanced operational efficiency. Traditional financial processes often involve manual, time-consuming tasks that are prone to errors. By implementing finance technologies, organisations can automate these processes, reducing the risk of human error and freeing up valuable time for finance professionals to focus on strategic initiatives. Additionally, finance technologies facilitate seamless collaboration and communication between different departments within an organisation, enabling smoother workflows and faster decision-making.

Finance technology plays a pivotal role in finance transformation by enabling organisations to transform their financial operations, enhance decision-making, and drive

DOI: 10.1201/9781003514503-7

strategic initiatives. These technologies encompass a wide range of tools and platforms that automate manual tasks, streamline processes, and provide real-time insights. For example, artificial intelligence and machine learning algorithms can analyse vast amounts of financial data, identify patterns, and predict future trends, enabling finance professionals to make data-driven decisions.

Moreover, finance technologies such as blockchain have the potential to revolutionise financial transactions by providing secure, transparent, and efficient methods for recording and verifying transactions. By leveraging blockchain technology, organisations can streamline payment processes, reduce transaction costs, and enhance the security and integrity of financial transactions.

Implementing finance technologies for finance transformation yields a multitude of benefits for organisations. Firstly, these technologies enable organisations to improve their financial planning and forecasting capabilities. By leveraging real-time data and advanced analytics, finance professionals can make accurate predictions and develop comprehensive financial plans that align with the organisation's strategic objectives.

Furthermore, finance technologies enhance risk management practices by providing organisations with real-time insights into potential risks and vulnerabilities. These technologies can identify anomalies, detect fraudulent activities, and highlight areas of potential financial risk, enabling organisations to take proactive measures to mitigate these risks.

Additionally, finance transformation through finance technologies enables organisations to enhance their operational efficiency by automating manual tasks and streamlining processes. This not only reduces the risk of errors but also frees up valuable time for finance professionals to focus on strategic initiatives and value-added activities.

While the benefits of finance transformation through finance technologies are undeniable, organisations may face certain challenges and obstacles during the implementation process. One of the primary challenges is the *resistance to change*. Finance professionals may be hesitant to embrace new technologies and may require proper training and education to fully leverage these tools. It is crucial for organisations to invest in change management initiatives and provide adequate support to ensure a smooth transition.

Another challenge is the *integration of finance technologies* with existing systems and processes. Organisations may have legacy systems that are not compatible with modern finance technologies, leading to complexities and delays in implementation. It is essential for organisations to conduct a thorough analysis of their existing systems and develop a comprehensive integration plan to ensure seamless adoption of finance technologies.

In my experience, in order to ensure successful finance transformation through the implementation of finance technologies, organisations should consider adopting the following best practices:

Define Clear Objectives: Organisations should clearly define their objectives and desired outcomes before embarking on a finance transformation journey. This will provide a roadmap for implementation and help measure the success of the transformation.

Engage Stakeholders: It is crucial to involve all relevant stakeholders throughout the transformation process. This includes finance professionals, IT teams, and senior management. Engaging stakeholders from the beginning ensures buy-in, promotes collaboration, and facilitates a smooth transition.

Invest in Training and Education: Proper training and education are essential to enable finance professionals to fully leverage finance technologies. Organisations should provide comprehensive training programmes and continuous learning opportunities to ensure that employees have the necessary skills and knowledge to utilise these technologies effectively.

Develop a Robust Change Management Strategy: Change management is vital to overcome resistance to change and ensure a smooth transition. Organisations should develop a robust change management strategy that includes effective communication, stakeholder engagement, and support mechanisms to facilitate the adoption of finance technologies.

Several companies have successfully implemented finance technologies to achieve finance transformation. One such case study is the *Volkswagen Group*, a global manufacturing company. Volkswagen Group implemented RPA in its finance department, automating manual tasks such as invoice processing and financial reporting. This resulted in significant time savings, reduced errors, and improved accuracy in financial reporting.

Another case study is *Lloyds Bank*, a financial services firm. Lloyds Bank adopted artificial intelligence and machine learning algorithms to analyse customer data and identify patterns in their spending habits. This enabled the company to personalise its financial products and services, resulting in increased customer satisfaction, retention, and revenue growth.

The future of finance technologies is promising, with several emerging trends set to revolutionise the financial landscape. One such trend is the increased adoption of cloud-based finance technologies. Cloud computing offers scalability, flexibility, and cost-efficiency, enabling organisations to access and analyse vast amounts of financial data in real-time.

Another emerging trend is the integration of finance technologies with other emerging technologies such as the Internet of Things (IoT) and big data analytics. This integration enables organisations to collect and analyse real-time financial data from various sources, providing valuable insights for decision-making.

Furthermore, the rise of decentralised finance (DeFi) and blockchain technology is set to transform financial transactions by eliminating intermediaries and enabling peer-to-peer transactions. This technology has the potential to disrupt traditional financial systems, making transactions faster, cheaper, and more secure.

Finance transformation through the adoption of emerging finance technologies is crucial for organisations aiming to stay competitive in today's business landscape. These technologies enable organisations to streamline financial processes, enhance decision-making capabilities, and drive strategic initiatives. While there may be challenges and obstacles in the implementation process, organisations can overcome them by following best practices and investing in change management initiatives.

The future of finance technologies holds immense potential, with emerging trends such as cloud computing, integration with other technologies, and the rise of DeFi set to reshape the financial landscape. Organisations that embrace these technologies and leverage their power will gain a competitive edge, enabling them to thrive in an increasingly digital and data-driven world. It is imperative for organisations to embrace finance technologies and embark on a finance transformation journey to secure their position in the evolving financial landscape.

CHAPTER 8

Finance Automation

IN TODAY'S FAST-PACED BUSINESS environment, companies are constantly seeking ways to stay ahead of the competition. One area that has seen significant advancements is finance automation. By leveraging technology and data-driven processes, finance teams can streamline their operations and add value to the organisation. Chief Finance Officer (CFO) must consider the role of finance automation in the digital transformation of businesses. We will discuss the benefits, challenges, and key players involved in this process.

The consistency principle is a fundamental tenet of financial reporting. It states that once an accounting method or principle has been chosen, it should be followed consistently throughout all accounting periods. However, the ever-changing business landscape necessitates a more agile approach to finance. By automating financial processes, finance teams can adapt quickly to changing circumstances and provide real-time insights to decision-makers.

Gone are the days when finance professionals could rely solely on traditional accounting skills. With the increasing reliance on technology and data, finance teams need to embrace automation and expand their skill sets. This shift requires a mindset of continuous learning and a willingness to explore new tools and technologies.

Siddiqui et al. (2023) contend automation will play a vital role in enabling digital transformation to take on the role of a change agent within the financial sector (Siddiqui et al., 2023). They argue that FinTechs are taking the lead in making best-practice use of automation to drive value generation and reduce operational costs.

The role of *coding* is vital in finance automation. While finance professionals may not need to become expert coders, having a basic understanding of coding can be invaluable in the era of finance automation. Learning coding languages like R and Python can enable finance teams to leverage data science techniques and extract actionable insights from vast amounts of financial data.

CFOs are not simply confined to number crunching and reporting. In the digital age, CFOs are expected to be strategic partners, leveraging technology and automation to drive business growth. The New Age CFO understands the importance of data and is proficient in coding languages and data analytics tools.

DOI: 10.1201/9781003514503-8

Cross-functional collaboration is important, as the success of finance automation relies on it. CFOs need to work closely with other senior leaders, such as the Chief Executive Officer (CEO), Chief Information Officer (CIO), Chief Operations Officer (COO), and Chief Data Officer (CDO), to align the organisation's goals and drive digital transformation. By speaking the language of developers and understanding the needs of different departments, CFOs can ensure the successful implementation of automation initiatives.

Managing change and overcoming resistance is critical. Implementing finance automation requires change management skills. CFOs need to act as change agents, rallying support from stakeholders and communicating the benefits of automation to the entire organisation.

Overcoming resistance to change and addressing concerns is crucial in ensuring a smooth transition to a more automated finance function.

An *effective finance automation team* can be a key success factor. An automation lead has a key role in acting as a project leader and overseeing the implementation of finance automation initiatives. This individual possesses a blend of finance and technical skills, enabling effective communication and coordination between different teams. The automation lead drives the automation strategy, identifies opportunities for process improvement, and ensures the successful integration of automation tools and technologies.

The *change agent* plays a critical role in driving the digital transformation of the finance function. This individual has a deep understanding of the company's culture, processes, and people. They communicate the vision and benefits of finance automation, build buy-in from stakeholders, and facilitate the change management process.

Collaborating with IT and Operations is an enabler. Finance automation requires collaboration with other departments, such as IT and operations. The CIO oversees the company's IT infrastructure and plays a vital role in integrating finance automation tools and systems. The COO focuses on operational efficiency and benefits from real-time financial data provided by finance automation.

Data scientists bring their expertise in data analysis and modelling to the finance automation team. They work closely with finance professionals to extract insights from financial data, develop predictive analytics models, and identify patterns and trends. Data scientists help finance teams make data-driven decisions and enhance financial forecasting and planning processes.

Benefits of automation include:

- *Improved Operational Efficiency:* Automation reduces manual tasks and streamlines finance processes, resulting in increased efficiency and productivity.

- *Real-Time Insights:* Finance automation provides real-time access to financial data, enabling faster decision-making and better financial forecasting.

- *Data Accuracy:* Automation minimises the risk of human errors and ensures data accuracy, leading to more reliable financial reports.

- *Cost Savings:* By automating routine tasks, finance teams can allocate resources more effectively, reducing costs and freeing up time for value-added activities.

Challenges of finance automation:

- *Resistance to Change:* Employees may resist the adoption of automation due to fear of job loss or unfamiliarity with new technologies.

- *Integration Complexity:* Integrating various systems and tools can be challenging, requiring collaboration between different departments and IT teams.

- *Data Security and Privacy:* Finance automation involves handling sensitive financial data, requiring robust security measures to protect against data breaches.

- *Skill Gap:* Implementing finance automation may require upskilling or hiring individuals with expertise in coding, data analytics, and automation tools.

Based on my experience, best practices in automation include:

Define Clear Objectives: Before embarking on finance automation, clearly define the objectives and expected outcomes. Identify the pain points in existing processes and determine how automation can address them. Set measurable goals and key performance indicators to track the success of automation initiatives.

Assess and Select Automation Tools: Conduct a thorough assessment of available automation tools and technologies. Consider factors such as scalability, integration capabilities, user-friendliness, and data security. Choose tools that align with the organisation's needs and are adaptable to future requirements.

A Comprehensive Automation Strategy: Developing a detailed automation strategy that outlines the roadmap for implementation. Identify the processes to be automated, prioritise them based on impact and feasibility, and create a timeline for implementation. Involve stakeholders from different departments to ensure alignment and buy-in.

Pilot and Iterate: Start with pilot projects to test the effectiveness of automation tools and processes. Gather feedback from end-users and stakeholders and iterate based on their inputs. Continuously monitor and evaluate the impact of automation, making adjustments as needed.

Training and Support: Invest in training programmes to upskill finance professionals and other team members involved in automation initiatives. Provide ongoing support and guidance to ensure a smooth transition to automated processes. Encourage a culture of continuous learning and innovation.

Finance automation is a crucial component of the digital transformation journey for businesses. By embracing automation, finance teams can streamline processes, enhance operational efficiency, and provide real-time insights to drive business growth.

The role of the New Age CFO is pivotal in leading this transformation, collaborating with other departments, and driving change management. By building an effective finance automation team and following best practices, organisations can harness the power of automation to stay ahead in today's competitive landscape. Embracing finance automation is not just a choice; it's a necessity for businesses aiming for long-term success in the digital era.

CHAPTER 9

Artificial Intelligence

FINANCE TRANSFORMATION ENABLES ORGANISATIONS to streamline their financial operations and adapt to changing business environments. With the rapid advancements in technology, one of the key drivers of finance transformation is artificial intelligence (AI). AI has the potential to revolutionise the way financial processes are carried out, providing organisations with greater efficiency, accuracy, and strategic insights.

AI, in simple terms, refers to the ability of machines to simulate human intelligence and perform tasks that typically require human intervention. In the context of finance, AI can be used to automate repetitive tasks, analyse large volumes of data, detect patterns, and make predictions. This enables finance professionals to focus on more value-added activities such as strategic planning, decision-making, and risk management.

An (2023) argues AI will play an increasingly pivotal role in finance transformation by automating mundane and time-consuming tasks, reducing errors, and improving the overall efficiency of financial processes (An, 2023). AI-powered tools can perform tasks such as data entry, reconciliations, and report generation with a higher degree of accuracy and speed than humans. This not only saves time but also eliminates the risk of human error, ensuring data integrity and reliability.

Moreover, AI algorithms can analyse vast amounts of financial data and identify patterns that may not be apparent to human analysts. By leveraging machine learning techniques, AI can make predictions and recommendations based on historical data, enabling organisations to optimise financial planning, budgeting, and forecasting processes. This empowers finance professionals to make data-driven decisions and drive strategic initiatives that contribute to the organisation's growth and profitability.

The use of AI in finance offers numerous benefits to organisations. Firstly, AI-driven automation reduces the manual effort required to complete financial tasks, allowing finance teams to focus on higher-value activities. This leads to increased productivity and efficiency, as well as cost savings in the long run.

Secondly, AI algorithms can analyse vast amounts of financial data in real-time, providing organisations with actionable insights and improving decision-making. For example, AI-powered fraud detection systems can identify anomalies in financial transactions and flag potentially fraudulent activities, enabling organisations to take proactive measures to mitigate risks.

Furthermore, AI can enhance customer experience by providing personalised financial advice and recommendations. AI-powered chatbots can interact with customers, answer their queries, and provide assistance in real-time, improving customer satisfaction and loyalty.

Qi (2023) contends that the use of AI will prove vital in transforming financial services over the next decade and lauds the use of robo-advisors, which are platforms offering automated financial services powered by AI algorithms (Qi, 2023). Qi argues robo-advisors bridge the gap between consumers and financial products and are more automated and intelligent than traditional investment advisors, significantly lowering labour expenses and enabling customers to do financial management from the comfort of their homes.

In my experience, whilst these claims sound tantalising, the reality is robo-advisers are not yet equipped to deal with market volatility. There is a high level of unpredictability in financial markets, so robo-advisers are more suited to an auxiliary role at present.

Several organisations have successfully harnessed the power of AI to transform their finance functions. One such case study is JP Morgan, which developed an AI-powered virtual assistant called Contract Intelligence (COIN). COIN can analyse legal documents and extract key information, reducing the time and effort required for contract review and analysis.

Another case study is PayPal, which uses AI algorithms to detect and prevent fraud. By analysing patterns in transaction data, PayPal's AI system can identify potentially fraudulent activities and take appropriate actions to protect its users.

Additionally, American Express utilises AI to provide personalised recommendations to its customers. By analysing transaction data and customer preferences, American Express can offer targeted offers and recommendations that align with individual customer needs and preferences.

These case studies demonstrate the transformative power of AI in finance and how organisations can leverage it to optimise their financial processes and deliver enhanced value to stakeholders.

While the benefits of AI in finance transformation are significant, there are several challenges that organisations need to overcome to successfully implement AI-powered solutions. One of the key challenges is the availability and quality of data. AI algorithms require large volumes of high-quality data to train and make accurate predictions. Organisations need to ensure that their data is clean, properly labelled, and accessible for AI systems to deliver meaningful insights.

Another challenge is the integration of AI into existing finance systems and processes. Many organisations have legacy systems that may not be compatible with AI technologies. The integration of AI requires careful planning, collaboration between IT and finance teams, and possibly the adoption of new technologies or systems.

Furthermore, there may be concerns about the ethical use of AI in finance. For example, AI systems that make lending decisions need to be fair and unbiased. Organisations need to ensure that AI systems are transparent, explainable, and auditable to address concerns around privacy, security, and compliance.

To successfully integrate AI into finance processes, organisations should follow a systematic approach. Firstly, it is important to identify the specific finance processes that can benefit from AI. This could include tasks such as data entry, invoice processing, financial analysis, or risk management.

Once the processes are identified, organisations should assess the readiness of their data. This involves cleaning and preparing the data, ensuring its accuracy and completeness, and making it accessible to AI systems.

Next, organisations should select the appropriate AI technologies and tools that align with their finance objectives. This could include machine learning algorithms, natural language processing (NLP), robotic process automation (RPA), or predictive analytics.

After selecting the tools, organisations should develop a proof-of-concept or pilot project to test the feasibility and effectiveness of AI in their finance processes. This allows for iterative improvements and fine-tuning of the AI models before full-scale implementation.

Finally, organisations should ensure ongoing monitoring and continuous improvement of AI systems. This involves regular evaluation of the performance, accuracy, and effectiveness of the AI models, as well as updating them with new data and evolving business requirements.

There are a variety of tools and technologies available that can help organisations harness the power of AI in finance. These include:

Machine Learning Algorithms: These algorithms enable machines to learn from data and make predictions or recommendations. They can be used for tasks such as fraud detection, credit scoring, or financial forecasting.

NLP: NLP enables machines to understand and interpret human language. NLP can be used to extract key information from financial documents, analyse customer feedback, or generate personalised financial advice.

RPA: RPA involves the use of software robots to automate repetitive tasks. In finance, RPA can be used for tasks such as data entry, invoice processing, or report generation, freeing up human resources for more strategic activities.

Predictive Analytics: Predictive analytics uses historical data and statistical models to make predictions about future outcomes. In finance, predictive analytics can be used for financial planning, budgeting, or risk management.

These tools and technologies provide organisations with the necessary capabilities to implement AI-driven solutions and transform their finance functions.

Several companies have successfully leveraged AI to transform their finance functions and achieve significant results. One such company is *HSBC*, which implemented an AI-powered system to automate its trade finance processes. The system reduced the time required for processing trade finance documents from several days to a few hours, improving operational efficiency and customer satisfaction.

Another case study is *BlackRock*, a global investment management company which uses AI algorithms to analyse market data and make investment decisions. The AI system can process vast amounts of data in real-time and identify investment opportunities or risks, enabling BlackRock to deliver superior investment performance to its clients.

Furthermore, Citigroup implemented an AI-powered system to automate its regulatory compliance processes. The system can analyse large volumes of financial data, identify potential compliance issues, and generate reports for regulatory authorities, ensuring timely and accurate compliance.

These case studies highlight how AI can be effectively utilised to transform finance processes, improve operational efficiency, and deliver tangible business benefits.

The future of finance transformation is closely intertwined with the continued advancements in AI. Some of the key trends that are expected to shape the future of finance include:

Augmented Analytics: Augmented analytics combines AI and human intelligence to provide more intuitive and interactive data analysis capabilities. This will enable finance professionals to uncover hidden insights, ask more sophisticated questions, and make more informed decisions.

Blockchain Technology: Blockchain has the potential to revolutionise financial transactions by providing secure, transparent, and decentralised systems. AI can complement blockchain by automating tasks such as smart contract execution, fraud detection, or identity verification.

Cognitive Robotic Process Automation (CRPA): CRPA combines RPA with cognitive technologies such as NLP and machine learning. This enables robots to perform complex tasks that require cognitive capabilities, such as analysing unstructured data or making intelligent decisions.

Explainable AI: As AI becomes more pervasive in finance, there is a growing need for transparency. Explainable AI systems provide clear explanations of how decisions are made, ensuring that AI models are fair, unbiased, and compliant with regulatory requirements.

These trends indicate that the future of finance transformation will be driven by AI technologies that enable organisations to leverage data, automate processes, and make better-informed decisions.

AI has the potential to revolutionise finance transformation by automating tasks, improving decision-making, and enhancing customer experience. Organisations that embrace AI in finance can streamline their processes, reduce costs, and gain a competitive edge in today's rapidly evolving business landscape.

While there are challenges in implementing AI, organisations can overcome them by following a systematic approach, leveraging the right tools and technologies, and ensuring ongoing monitoring and improvement. The success stories of companies that have already transformed their finance functions using AI serve as inspiration and guidance for others.

As we look to the future, the integration of AI into finance processes will continue to evolve, driven by advancements in technology and changing business needs. By harnessing the power of AI, organisations can unlock new opportunities, drive innovation, and achieve transformational results in finance. It is an exciting time for the future of finance, and with AI as a catalyst, the possibilities are limitless.

AI has become a game-changer in the finance industry, revolutionising the way businesses operate and transforming traditional financial processes. By harnessing the power of AI, organisations can unlock valuable insights from their data, enabling them to make informed decisions and drive growth. CFOs must consider the world of finance transformation through the lens of AI and explore several compelling case studies that highlight the significant impact this technology can have on financial forecasting, fraud detection, risk assessment, and process automation.

9.1 AI-IMPROVED FINANCIAL FORECASTING AND BUDGETING: IBM

One of the key challenges that finance professionals face is accurately predicting future financial trends and preparing budgets accordingly. Traditional forecasting methods often fall short, relying heavily on historical data and human intuition. However, with AI, organisations can leverage advanced algorithms and machine learning techniques to analyse vast amounts of data from various sources, including market trends, customer behaviour, and economic indicators.

A prime example of AI-powered financial forecasting and budgeting is demonstrated by IBM. IBM was an early pioneer in developing AI with its planning and analytics bot called Watson. Watson's AI model offers sophisticated forecasting that gives businesses the ability to anticipate future performance more accurately as well as make better-informed, data-driven decisions. By implementing an AI-based predictive analytics system, they were able to analyse historical financial data, market trends, and external factors to generate highly accurate forecasts. This allowed them to make informed decisions regarding resource allocation, investment strategies, and risk management. Consequently, IBM experienced a substantial increase in profitability and efficiency, setting a new standard in financial forecasting.

9.2 AI FOR FRAUD DETECTION: MASTERCARD AND CITIBANK

Fraud detection is a paramount concern for the finance industry, as fraudulent activities can have severe financial implications and damage an organisation's reputation. Traditional fraud detection methods often rely on rule-based systems that are limited in their ability to adapt to evolving fraudulent techniques. AI, on the other hand, offers a more proactive and effective approach to identifying and preventing fraudulent activities.

Mastercard, a leading financial services organisation, successfully implemented AI-based fraud detection systems to combat financial fraud. Mastercard's Decision Intelligence technology uses patterns from historical shopping and spending habits of cardholders to set a behavioural baseline against which it compares each new transaction. Mastercard is selling an AI tool to protect banks. The 'Consumer Fraud Risk' solution, now live in the United Kingdom, has been adopted by nine of the biggest British banks, including Lloyds Banking Group Plc, Natwest Group Plc, and Bank of Scotland Plc.

Through Citi Ventures, CitiBank made a strategic investment in Feedzai, a global data science enterprise that works in real-time to identify and eradicate financial crimes such as fraud. Through its continuous and rapid evaluation of large amounts of data, Feedzai conducts large-scale analyses to identify fraudulent or questionable activity and alert the customer. By analysing vast volumes of transactional data and applying machine learning algorithms, these financial organisations were able to detect patterns and anomalies indicative of fraudulent behaviour. This allowed them to identify suspicious transactions in real-time, significantly reducing the risk of financial loss and protecting their customers' assets. The AI-powered fraud detection system not only enhanced their security measures but also improved customer trust and satisfaction.

9.3 AI-POWERED RISK ASSESSMENT AND MITIGATION: SWISS RE

Effective risk assessment and mitigation are crucial components of financial management, enabling organisations to identify potential risks, develop appropriate strategies, and safeguard their financial well-being. AI offers a powerful tool to enhance risk assessment processes, providing organisations with real-time insights and predictive analytics to make proactive risk management decisions.

Swiss Re Insurance, a leading insurance provider, integrated AI technology into their risk assessment framework. By analysing vast amounts of customer data, market trends, and historical claims data, the company was able to develop a sophisticated risk model that accurately predicted the likelihood of claims and estimated potential losses. This AI-powered risk assessment system allowed XYZ Insurance to optimise their underwriting processes, reduce fraudulent claims, and improve overall profitability. The integration of AI not only transformed their risk management practices but also enabled them to provide more competitive and tailored insurance products to their customers.

9.4 AUTOMATING FINANCIAL PROCESSES WITH AI: JP MORGAN CHASE

Finance departments are often burdened with repetitive and time-consuming tasks, such as data entry, reconciliation, and reporting. Manual processing not only slows down operations but also increases the risk of errors. AI has the potential to automate these processes, freeing up valuable time and resources for finance professionals to focus on more strategic and value-added activities.

JP Morgan Chase has been successfully leveraging RPA for a while now to perform tasks such as extracting data, comply with Know Your Customer regulations, and capture documents. Like many other financial organisations, they are embracing AI to automate their financial processes. By leveraging RPA and NLP capabilities, they were able to streamline data entry, automate reconciliation, and generate real-time financial reports. This significantly reduced processing time, eliminated errors, and enhanced overall efficiency. The implementation of AI not only improved operational effectiveness but also enabled JP Morgan Chase to allocate their resources more effectively, driving innovation and growth.

9.5 ROBO-ADVISORS WITH AI: VANGUARD

Robo-advisors are another case study. These are digital platforms that provide automated, algorithm-driven financial planning services with minimal human supervision. Unlike their human counterparts, robo-advisors monitor the markets non-stop, 24 hours a day, seven days a week. Robo-advisors can also offer investors up to 70% in cost savings and typically require lower or no minimums to participate. Many large investment firms have their own robo-advisor solutions, such as FidelityGo and Schwab's Intelligent.

Portfolio and Vanguard's personal advisor services robot. There are also several startups vying for the spoils, such as Betterment, wealthfront, and SoFi, in this already crowded space.

While AI offers immense potential for finance transformation, organisations need to address several challenges and considerations when implementing this technology. Firstly, data quality and availability are crucial for AI algorithms to generate accurate insights.

Organisations must ensure that their data is clean, consistent, and readily accessible. Secondly, there is a need for skilled AI professionals who can develop and maintain AI systems. Organisations should invest in training and hiring experts in AI technology to ensure successful implementation. Lastly, ethical considerations such as data privacy and security should be taken into account to maintain customer trust and regulatory compliance.

The adoption of AI in finance transformation provides numerous benefits to organisations. Firstly, AI enables faster and more accurate decision-making by analysing vast amounts of data and identifying patterns and trends. This empowers finance professionals to make informed decisions, optimise resource allocation, and mitigate risks effectively. Secondly, AI automates repetitive tasks, reducing manual effort and freeing up valuable time for finance professionals to focus on strategic initiatives. Thirdly, AI improves operational efficiency by eliminating errors, streamlining processes, and enhancing overall productivity. Lastly, AI enhances customer experience by enabling personalised financial services, fraud detection, and tailored insurance offerings.

As AI technology continues to evolve, several exciting trends and developments are emerging in the field of finance transformation. Firstly, NLP and machine learning algorithms are becoming more sophisticated, enabling AI systems to understand complex financial documents, contracts, and regulations. This will significantly streamline compliance processes and enhance regulatory reporting. Secondly, AI-powered chatbots and virtual assistants are revolutionising customer service in the finance industry. These intelligent systems can provide personalised financial advice, answer queries, and facilitate seamless transactions. Lastly, AI is facilitating the integration of blockchain technology into finance processes, enabling secure and transparent transactions, smart contracts, and decentralised finance.

AI has undoubtedly transformed the finance industry, offering unprecedented opportunities for finance professionals to extract valuable insights from their data and drive strategic decision-making. Through case studies in financial forecasting, fraud detection, risk assessment, and process automation, we have witnessed the significant impact AI can have on finance transformation. While challenges and considerations exist, the benefits of AI are far-reaching, enhancing decision-making, automating processes, and improving operational efficiency. As we look to the future, AI will continue to shape the finance industry, revolutionising customer experience, regulatory compliance, and the overall landscape of financial services. Embracing AI is no longer an option but a necessity for organisations seeking to thrive in the ever-evolving world of finance transformation.

CHAPTER 10

Machine Learning

MACHINE LEARNING (ML) IS a branch of artificial intelligence (AI) that focuses on developing algorithms and models that enable computers to learn and make predictions or decisions without being explicitly programmed. In other words, it is about creating systems that can automatically learn from data and improve their performance over time. ML algorithms can analyse large amounts of data, identify patterns, and make predictions or decisions based on those patterns.

Ilsøe et al. (2022) argue that since the Great Recession of 2008, new digital technologies and ML in particular are seen as an enabler for disruption (Ilsøe et al., 2022). Challenger businesses and FinTechs are using ML to offer 'one-stop-shop' business models that provide all the services a customer would require. FinTechs are seen as challengers forging new types of financial businesses by providing updated, mobile, and innovative digital services, exploiting the possibilities of app-based services, ML and AI.

Kamuangu et al. (2024) contend that ML will open up new opportunities in financial services in the fields of fraud detection, risk management, customer service, personalisation, predictive analytics, and supporting decision-making. These are explored in greater depth later in the chapter. They also accept the ethical, regulatory, privacy, and security implications that these ML applications bring (Kamuangu and K K, 2024).

In the world of finance, where vast amounts of data are generated and processed every day, ML has emerged as a powerful tool for driving transformation. By leveraging ML algorithms, financial institutions can gain valuable insights from their data, automate processes, and make more informed decisions.

ML can be applied to various aspects of finance, such as risk assessment, fraud detection, portfolio optimisation, and customer segmentation. It can analyse historical data to identify patterns and trends, which can then be used to predict future market behaviour or customer preferences. This predictive capability can help financial institutions optimise their operations, improve risk management, and enhance customer experience.

In my experience, the use of ML in finance offers several benefits. Firstly, it enables financial institutions to process and analyse vast amounts of data in real-time, which can lead to faster and more accurate decision-making. This can be particularly valuable

DOI: 10.1201/9781003514503-10

in high-frequency trading or risk management, where timely and accurate information is crucial.

Secondly, ML algorithms can identify complex patterns and relationships in data that may not be apparent to human analysts. This can uncover valuable insights and help financial institutions make more informed decisions. For example, ML algorithms can analyse historical market data to identify hidden correlations between different asset classes, which can then be used to optimise portfolio allocation.

Finally, ML can automate repetitive tasks and improve operational efficiency. By automating processes such as data entry, data cleansing, and data analysis, financial institutions can free up valuable resources and focus on more strategic activities. This can lead to cost savings and increased productivity.

ML has already found numerous applications in the world of finance. One such application is credit scoring, where ML algorithms are used to predict the creditworthiness of borrowers. By analysing various data points such as income, employment history, and credit history, ML algorithms can assess the risk associated with lending to a particular individual or business.

Another application of ML in finance is fraud detection. ML algorithms can analyse large volumes of transactional data in real-time to identify suspicious patterns or anomalies that may indicate fraudulent activity. By quickly flagging potentially fraudulent transactions, financial institutions can take appropriate action to prevent financial loss and protect their customers.

ML also plays a significant role in algorithmic trading, where complex trading strategies are executed using pre-defined rules and models. ML algorithms can analyse historical market data, identify patterns, and generate trading signals. These signals can then be used to automate the execution of trades, leading to faster and more efficient trading.

Several case studies highlight the transformative impact of ML in finance. One such case study is *JP Morgan* and their use of ML algorithms to analyse legal documents. By automating the analysis of legal documents such as loan agreements, JP Morgan was able to reduce the time and resources required for this task, improving efficiency and reducing costs.

Another case study is *PayPal* and its use of ML for fraud detection. By leveraging ML algorithms to analyse transactional data, PayPal was able to reduce fraud losses by 25%, resulting in significant financial savings.

Goldman Sachs is another example of a financial institution that has embraced ML. They have developed ML models that analyse market data to identify trading opportunities, resulting in improved trading performance and increased profitability.

These case studies demonstrate how ML can drive finance transformation by enabling financial institutions to automate processes, make more informed decisions, and improve operational efficiency.

While ML offers significant benefits, it also presents challenges and limitations that need to be considered. One challenge is the need for high-quality and reliable data.

ML algorithms rely on large amounts of data to learn and make accurate predictions. Therefore, financial institutions need to ensure that their data is clean, consistent, and representative of the problem they are trying to solve.

Another challenge is the interpretability of ML models. Some ML algorithms, such as deep learning neural networks, can be highly complex and difficult to interpret. This can make it challenging for financial institutions to explain the reasoning behind their decisions, which can be a regulatory requirement or a customer expectation.

Furthermore, ML models are not infallible and can make mistakes. Financial institutions need to carefully evaluate the performance of their ML models and have appropriate mechanisms in place to monitor and mitigate potential risks.

Implementing ML in finance transformation requires a systematic approach. Financial institutions should start by identifying the business problems or opportunities where ML can add value. This could involve assessing existing processes, identifying pain points, and understanding how ML can address those challenges.

Once the problem or opportunity has been identified, financial institutions need to gather and prepare the relevant data. This may involve aggregating data from various sources, cleaning and transforming the data, and ensuring its quality and integrity.

Next, financial institutions should select and develop appropriate ML models. This may involve selecting the appropriate algorithm, training the model on historical data, and validating its performance using suitable evaluation metrics.

Finally, financial institutions need to integrate the ML models into their existing systems and processes. This may involve developing APIs or interfaces to enable seamless integration with other systems and providing appropriate training and support to ensure successful adoption.

To ensure successful implementation of ML in finance transformation, financial institutions should follow some best practices:

Start Small and Scale: Begin with a small pilot project to test the feasibility and value of ML in a controlled environment. Once the pilot project has been successful, financial institutions can scale up and extend the use of ML to other areas.

Invest in Data Quality and Infrastructure: High-quality data and robust infrastructure are essential for successful ML implementation. Financial institutions should invest in data governance, data management, and infrastructure to ensure the availability, reliability, and security of data.

Foster Collaboration between Data Scientists and Domain Experts: Successful ML implementation requires collaboration between data scientists, who have the technical expertise, and domain experts, who have the business knowledge. By working together, they can develop ML models that are both technically sound and aligned with business objectives.

Continuously Monitor and Evaluate Performance: ML models need to be continuously monitored and evaluated to ensure their performance remains optimal. Financial institutions should establish processes for monitoring model performance, detecting and mitigating risks, and updating models as required.

The field of ML is continuously evolving, and there are several future trends and advancements that hold promise for finance transformation.

One such trend is the increasing use of deep learning algorithms. Deep learning is a subset of ML that focuses on training artificial neural networks with multiple layers.

Deep learning algorithms have shown significant potential in areas such as image recognition and natural language processing. In finance, deep learning algorithms can be used for tasks such as sentiment analysis, fraud detection, and algorithmic trading.

Another trend is the use of reinforcement learning in finance. Reinforcement learning is a type of ML that involves an agent learning how to interact with an environment to maximise a reward. In finance, reinforcement learning can be used to develop trading strategies that adapt and improve over time based on market conditions and feedback.

Furthermore, advancements in data analytics and cloud computing are enabling financial institutions to process and analyse even larger volumes of data, leading to more accurate and sophisticated ML models.

ML has the potential to drive significant transformations in the field of finance. By leveraging ML algorithms, financial institutions can gain valuable insights from their data, automate processes, and make more informed decisions. However, implementing ML in finance transformation requires careful planning, data management, and collaboration between data scientists and domain experts. By embracing ML, financial institutions can unlock new opportunities, improve operational efficiency, and stay ahead in an increasingly competitive landscape.

ML has emerged as a powerful tool for driving finance transformation. By leveraging ML algorithms, financial institutions can gain valuable insights from their data, automate processes, and make more informed decisions. The benefits of using ML in finance are numerous, including faster and more accurate decision-making, uncovering valuable insights, and improving operational efficiency.

Real-world applications of ML in finance include credit scoring, fraud detection, and algorithmic trading. These applications have already shown significant results, such as reducing fraud losses and improving trading performance.

However, implementing ML in finance transformation does come with its challenges and limitations. Financial institutions need to ensure high-quality and reliable data, address the interpretability of ML models, and carefully evaluate and mitigate potential risks.

To successfully implement ML in finance, financial institutions should follow best practices such as starting small and scaling up, investing in data quality and infrastructure, fostering collaboration between data scientists and domain experts, and continuously monitoring and evaluating performance.

Looking to the future, advancements in deep learning, reinforcement learning, and data analytics hold promise for further enhancing the role of ML in finance transformation.

By embracing ML, financial institutions can unlock new opportunities, improve operational efficiency, and stay ahead in an increasingly competitive landscape. It is an exciting time for finance transformation, and ML is at the heart of it.

CHAPTER 11

ERP Systems and Implementation

Finance transformation is a crucial aspect of any organisation's growth and success. It involves re-evaluating and optimising financial processes to enhance efficiency, reduce costs, and improve decision-making. One of the key drivers of finance transformation is the implementation of enterprise resource planning (ERP) systems. ERP systems provide a comprehensive and integrated solution for managing various financial activities, such as accounting, budgeting, financial reporting, and cash management.

Olaoye et al. (2024) argue ERP systems play a crucial role in facilitating and enabling digital transformation initiatives. They provide a centralised platform for managing core business functions such as finance, human resources, supply chain, manufacturing, and customer relationship management. By leveraging ERP, organisations can harness the power of digital technologies to drive innovation, improve efficiency, and deliver exceptional customer experiences (Olaoye and Daniel, 2024).

ERP systems play a vital role in finance transformation by streamlining financial processes and providing real-time visibility into financial data. These systems consolidate data from different departments and automate manual tasks, reducing the risk of errors and improving data accuracy. With ERP systems, finance teams can access critical financial information anytime, anywhere, enabling them to make informed decisions quickly. Moreover, ERP systems facilitate standardisation of financial processes across the organisation, ensuring consistency and compliance with regulatory requirements.

In my experience, implementing ERP systems in finance processes offers several benefits that contribute to finance transformation. Firstly, it improves operational efficiency by automating repetitive tasks, such as data entry and reconciliation, allowing finance professionals to focus on value-added activities. This automation also reduces the dependency on manual processes, minimising the risk of errors and delays. Additionally, ERP systems provide robust financial reporting capabilities, generating accurate and timely financial statements, which, in turn, enhance decision-making and enable better financial planning and forecasting.

DOI: 10.1201/9781003514503-11

While implementing ERP systems in finance processes can bring significant benefits, it is important to be aware of the challenges and considerations involved. One major challenge is the complexity of ERP system implementation. It requires careful planning, extensive data migration, and integration with existing systems. Organisations need to allocate sufficient time, resources, and expertise to ensure a smooth implementation process.

Another consideration is the cost associated with ERP system implementation. Apart from the initial investment in software licenses and hardware infrastructure, there are additional costs for customisation, training, and ongoing maintenance. Organisations need to carefully evaluate the return on investment and assess the long-term benefits of implementing an ERP system for finance transformation.

Furthermore, change management is a critical aspect of ERP system implementation. Finance professionals and other stakeholders should be involved throughout the process to ensure their buy-in and smooth transition. Adequate training and support should be provided to help employees understand the new system and its functionalities. Effective change management strategies can mitigate resistance to change and foster a positive attitude towards finance transformation.

Gunturu et al. (2024) contend modern ERP incorporates cloud computing, artificial intelligence, and data analytics technologies to facilitate automation in ERP payment systems (Gunturu et al., 2024).

To successfully implement ERP systems for finance transformation, organisations should follow a structured approach. The following steps can guide the implementation process:

Assess Finance Processes: Before implementing an ERP system, it is essential to evaluate existing finance processes. Identify pain points, inefficiencies, and areas for improvement. This assessment will help in determining the requirements and objectives of the ERP system.

Define Goals and Objectives: Clearly define the goals and objectives of implementing an ERP system for finance transformation. These goals can include improving financial reporting accuracy, reducing processing time, enhancing data visibility, or streamlining budgeting and forecasting processes.

Select the Right ERP System: Conduct thorough research and evaluate the different ERP systems available on the market. Consider factors such as scalability, flexibility, user-friendliness, and compatibility with existing systems. Choose an ERP system that aligns with the organisation's requirements and long-term goals.

Plan and Prepare: Develop a detailed implementation plan, considering timelines, milestones, and resource allocation. Create a project team comprising finance professionals, IT experts, and ERP system implementation specialists. Define roles and responsibilities, and establish communication channels for effective collaboration.

Data Migration and Integration: Data migration is a critical step in ERP system implementation. Ensure that data from existing systems is accurately migrated to the new ERP system. Also, integrate the ERP system with other financial tools, such as payroll and accounts payable systems, to ensure seamless data flow.

Testing and Training: Conduct thorough testing of the ERP system to ensure its functionality and compatibility with different finance processes. Provide comprehensive training to finance professionals and end-users to familiarise them with the new system and its features. This training should cover data entry, reporting, and other key functionalities.

Go-Live and Continuous Improvement: Once the ERP system is tested and the users are trained, it is time to go live. Monitor the system closely during the initial period and address any issues that arise. Continuous improvement should be a priority, with regular reviews and updates to optimise the system's performance and address emerging needs.

Several organisations have successfully transformed their finance processes through the implementation of ERP systems. Let's explore a couple of case studies that highlight the benefits and outcomes of such transformations.

Ford: Ford, a multinational manufacturing company, implemented an ERP system to streamline its finance processes. The system integrated various financial activities, such as accounts receivable, accounts payable, and financial reporting. As a result, the company experienced improved data accuracy, reduced processing time, and enhanced financial visibility. The finance team could generate real-time reports, enabling them to analyse financial performance and make data-driven decisions. The implementation of the ERP system also facilitated standardisation of financial processes across the organisation, ensuring consistency and compliance.

Walmart: Walmart, a retail organisation, implemented an ERP system to automate its budgeting and forecasting processes. The system allowed the finance team to create dynamic budgets, track expenses, and generate accurate financial forecasts. By automating these processes, it reduced the time and effort required for budgeting, enabling finance professionals to focus on strategic financial analysis. The ERP system also provided real-time visibility into financial data, allowing the finance team to identify cost-saving opportunities and make informed decisions.

ERP systems offer a wide range of features and functionalities that drive finance transformation. Some key features include:

General Ledger: ERP systems provide a centralised and automated general ledger that captures all financial transactions. This ledger forms the foundation for financial reporting and analysis.

Accounts Payable and Receivable: ERP systems automate the accounts payable and receivable processes, including invoice management, payment processing, and reconciliation. This automation reduces manual errors and improves cash flow management.

Financial Reporting: ERP systems offer robust reporting capabilities, enabling finance professionals to generate accurate and timely financial statements. These reports provide critical insights for decision-making and compliance purposes.

Budgeting and Forecasting: ERP systems facilitate budgeting and forecasting processes by automating data collection, analysis, and consolidation. This automation improves the accuracy and timeliness of financial forecasts.

Cash Management: ERP systems provide tools for effective cash management, including cash flow forecasting, bank reconciliation, and treasury management. These features help organisations optimise cash utilisation and mitigate financial risks.

Effective training and support are essential for successful ERP system implementation in finance processes. Organisations should provide comprehensive training programmes to finance professionals and end-users to ensure they are proficient in using the new system. This training should cover various aspects, such as data entry, reporting, and system navigation.

Additionally, organisations should establish a support system to address any issues or concerns that arise during and after the implementation process. This support system can include a dedicated help desk, online resources, and regular communication channels for users to seek assistance and provide feedback. Continuous training and support are crucial for maximising the benefits of ERP system implementation and driving finance transformation.

Integrating ERP systems with other financial tools is crucial for seamless data flow and efficient financial processes. Here are some best practices for successful integration:

Evaluate Compatibility: Before integrating an ERP system with other financial tools, evaluate their compatibility and ensure they can exchange data seamlessly. Consider factors such as data formats, APIs, and data synchronisation capabilities.

Define Data Mapping: Clearly define the mapping of data fields between the ERP system and other financial tools. This mapping should ensure that data is transferred accurately and consistently across systems.

Establish Data Governance: Implement data governance policies and procedures to maintain data integrity and consistency across integrated systems. Define data ownership, data validation rules, and data management processes to ensure accurate and reliable data.

Regularly Monitor and Test: Regularly monitor the integration between the ERP system and other financial tools to ensure data accuracy and system performance. Conduct periodic testing to identify and address any issues or discrepancies.

Ensure Security and Compliance: Implement robust security measures to protect data during integration. Ensure compliance with data protection regulations and industry standards, such as GDPR or HIPAA, depending on the organisation's jurisdiction and industry.

ERP systems have become a game-changer in finance transformation, enabling organisations to streamline financial processes, enhance data visibility, and drive informed decision-making. With their comprehensive features and functionalities, ERP systems provide a holistic solution for managing various finance activities. However, successful implementation requires careful planning, evaluation of challenges, and effective change management. By following best practices and leveraging the benefits of ERP systems, organisations can achieve significant finance transformation and position themselves for future success.

CHAPTER 12

Robotic Process Automation (RPA)

Eliminating Redundancy and Maximising Capacity

A S BUSINESSES STRIVE FOR success and growth, it becomes crucial to identify and eliminate areas of waste and redundancy. In the finance department, these inefficiencies can lead to significant losses in both time and resources. Common areas of waste in finance include manual data entry, outdated processes, and excessive paperwork. Redundancy often arises from duplicated efforts, such as multiple teams performing similar tasks or using multiple systems that lack integration.

An (2023) argues areas of waste and redundancy not only hinder the efficiency of the finance department but also impact the entire organisation, so robotic process automation (RPA) and artificial intelligence (AI) are vital to reduce labour costs and improve accuracy (An, 2023). Time spent on manual tasks could be better utilised in strategic decision-making and analysis. Moreover, redundant processes lead to increased costs and decreased productivity. To overcome these challenges, finance transformation emerges as a powerful solution.

Finance transformation is a comprehensive approach to reshape and optimise financial processes and systems. By leveraging technology, automation, and streamlined workflows, organisations can revolutionise their finance department and drive efficiency. The impact of finance transformation extends far beyond the finance team. It influences decision-making, enhances collaboration, and empowers organisations to adapt to changing market dynamics.

Through finance transformation, businesses can achieve increased accuracy in financial reporting and forecasting. Automated processes reduce the risk of errors that may arise from manual data entry and repetitive tasks. Additionally, finance transformation enables real-time data analysis, providing decision-makers with timely insights to make informed

DOI: 10.1201/9781003514503-12

choices. By harnessing technology and eliminating redundancy, organisations can focus on value-adding activities and drive growth.

Reducing waste and redundancy is an important component of transformation. Finance transformation offers a roadmap to reduce waste and redundancy in the finance department. One of the key steps is to identify and eliminate manual processes. Automation tools, such as RPA, can be implemented to streamline tasks like data entry, reconciliation, and reporting. By leveraging RPA, businesses can significantly reduce errors, enhance accuracy, and free up capacity for more strategic activities.

Another critical aspect of finance transformation is the integration of systems and processes. Often, organisations rely on multiple disconnected systems for different finance functions. This leads to redundant efforts and data inconsistencies. By integrating these systems, businesses can eliminate duplication, improve data accuracy, and simplify processes.

Integration also enables real-time visibility into financial data, empowering decision-makers to respond swiftly to market changes.

Maximising capacity is crucial. Finance transformation not only eliminates redundancy but also frees up capacity within the finance department. By automating repetitive tasks, finance professionals can redirect their efforts towards value-adding activities. This includes financial analysis, strategic planning, and business partnering. By maximising capacity, finance teams can contribute more effectively to the overall growth and success of the organisation.

Moreover, finance transformation enables organisations to reassess their RPA programs. While RPA brings significant benefits, it is essential to periodically review its implementation to ensure optimal outcomes. By leveraging the power of finance transformation, businesses can identify areas where RPA can be further enhanced or expanded. This proactive approach allows organisations to stay ahead of the curve and continually improve efficiency.

In my experience, successful finance transformation requires a well-defined strategy. Here are some key strategies to consider:

Assess Current Processes: Conduct a thorough evaluation of existing finance processes to identify areas of waste and redundancy.

Set Clear Objectives: Define specific objectives and outcomes that finance transformation aims to achieve.

Develop a Roadmap: Create a detailed plan outlining the steps, timeline, and resources required for successful finance transformation.

Embrace Automation: Leverage automation tools, such as RPA and intelligent process automation (IPA), to streamline finance processes and reduce manual efforts.

Promote Collaboration: Foster collaboration between finance and other departments to ensure alignment and integration of processes.

Invest in Training and Development: Equip finance professionals with the necessary skills and knowledge to adapt to the changing landscape of finance transformation.

Monitor and Measure: Establish key performance indicators (KPIs) to track the progress and effectiveness of finance transformation initiatives.

Implementing finance transformation requires careful planning and execution. Here are some steps to guide you:

Gain Leadership Support: Secure buy-in from top management to ensure commitment and allocation of necessary resources.

Assemble a Dedicated Team: Form a team with members from finance, IT, and other relevant departments to drive finance transformation.

Map Out Processes: Document current finance processes and identify areas for improvement.

Identify Technology Solutions: Research and select appropriate technology solutions that align with your finance transformation objectives.

Develop a Change Management Plan: Implement change management strategies to address any resistance or challenges that may arise during the transformation process.

Pilot and Test: Conduct pilot projects to test the effectiveness of new processes and technologies before full-scale implementation.

Continuously Evaluate and Adapt: Regularly assess the impact of finance transformation and make necessary adjustments based on feedback and results.

Viswanathan (2022) argues RPA technology can be used extensively to automate transaction processing and communication across various systems and RPA case studies include use for

1. Treasury processes, budgeting, planning, and forecasting

2. Billings and accounting

3. Inter-company transactions, allocations and adjustments, and journal entry

4. Reporting (internal and external finances) (Viswanathan, 2022)

Numerous organisations have successfully implemented finance transformation initiatives with RPA. Let's explore two case studies that highlight the benefits of finance transformation:

Munich Re: By implementing finance transformation, Munich Re reduced manual data entry by 70%, enabling finance professionals to focus on strategic analysis. This resulted in improved financial forecasting accuracy and faster decision-making.

The Home Depot: Through finance transformation, The Home Depot integrated their finance systems, eliminating redundant processes and ensuring real-time data visibility. This led to significant cost savings and improved collaboration between finance and other departments.

These case studies demonstrate the positive impact finance transformation can have on organisations of various sizes and industries.

Finance transformation relies on the effective utilisation of tools and technologies. By leveraging these tools and technologies, organisations can unlock the full potential of finance transformation. Here are some essential tools to consider:

RPA: Automates repetitive manual tasks, such as data entry and reconciliation, freeing up capacity and reducing errors.

IPA: Combines RPA with machine learning and AI to handle complex finance processes and decision-making.

Cloud-Based Financial Management Systems: Enables real-time data access, collaboration, and scalability.

Data Analytics and Visualisation Tools: Provides insights and visual representation of financial data for better decision-making.

Workflow Automation Software: Streamlines and automates approval processes, reducing bottlenecks and improving efficiency.

Finance transformation offers numerous benefits across different industries. These benefits demonstrate the wide-ranging impact that finance transformation can have on diverse industries. Let's explore some key advantages:

Manufacturing: Improved inventory management, cost control, and supply chain visibility.

Retail: Enhanced forecasting accuracy, inventory optimisation, and seamless omni-channel integration.

Healthcare: Streamlined revenue cycle management, improved billing accuracy, and enhanced compliance.

Technology: Real-time financial data for agile decision-making, improved profitability analysis, and scalability.

Services: Efficient project cost management, streamlined billing processes, and enhanced customer profitability analysis.

In today's rapidly evolving business landscape, finance transformation is no longer a luxury but a necessity. By eliminating waste and redundancy, organisations can unlock their full potential and maximise capacity within the finance department. Successful finance transformation requires a strategic approach, leveraging automation, integration, and collaboration. With the right tools and technologies, businesses can achieve enhanced accuracy, real-time insights, and improved decision-making.

As we move towards a future where efficiency and agility are paramount, embracing finance transformation becomes imperative. By reassessing RPA programmes, streamlining processes, and leveraging technology, organisations can position themselves for sustainable growth and success. Invest in finance transformation today and unlock the power to eliminate redundancy and maximise capacity in your organisation.

CHAPTER 13

Accelerating Cloud Adoption

IN TODAY'S RAPIDLY EVOLVING business landscape, finance departments are under increasing pressure to transform and adapt to new technologies. One technology that has gained significant traction in recent years is cloud computing. The cloud offers a multitude of benefits for finance transformation, including improved scalability, cost savings, enhanced security, and streamlined processes.

Viswanathan (2022) argues cloud adoption has brought acceleration into financial services. They argue that cloud-enabled applications provide increased security and scalability to the entire system for critical functions like consumer payments, credit scoring, statements, and billings. The cloud's intrinsic features, such as resource pooling, availability, on-demand service, security, and easy maintenance, are the primary reasons for its growth and popularity (Viswanathan, 2022).

Cloud technology enables finance teams to access their data and applications from anywhere, at any time, using any device with an internet connection. This flexibility is particularly valuable for organisations with multiple locations or remote workers. By centralising financial data in the cloud, finance professionals can collaborate more effectively and make better-informed decisions.

Finance often shifts key processes to the cloud, but it needs to demonstrate value in scaling processes. Cloud-based solutions also offer enhanced scalability, allowing finance departments to easily adjust their computing resources based on their needs. This agility is crucial for organisations experiencing rapid growth or seasonal fluctuations in demand. With the cloud, finance teams can quickly scale up or down their computing power, ensuring optimal performance and cost efficiency.

Moreover, cloud adoption can lead to significant cost savings for finance departments. By eliminating the need for on-premises infrastructure and reducing maintenance costs, organisations can allocate their resources to more strategic initiatives. Cloud-based solutions also offer subscription-based pricing models, allowing finance teams to pay only for the services they use, further optimising their costs.

Accelerating cloud adoption in finance can bring numerous benefits to organisations. Firstly, it allows for faster deployment of new applications and technologies. With traditional

DOI: 10.1201/9781003514503-13

on-premises solutions, implementing new software or upgrading existing systems can be time-consuming and costly. However, with the cloud, finance departments can leverage pre-built solutions and easily integrate them into their existing infrastructure, accelerating the deployment process.

Additionally, cloud technology enables finance teams to leverage advanced analytics and artificial intelligence tools more effectively. By harnessing the power of the cloud, organisations can access vast amounts of data and gain valuable insights to drive financial performance. Cloud-based analytics platforms provide real-time reporting and predictive analytics capabilities, empowering finance professionals to make data-driven decisions and identify new growth opportunities.

Another key benefit of accelerating cloud adoption in finance is enhanced security and data protection. Cloud service providers invest heavily in robust security measures and compliance frameworks to ensure the safety of their customers' data. By leveraging the expertise and resources of cloud providers, finance departments can enhance their data security and reduce the risk of cyber threats and data breaches.

Furthermore, cloud-based solutions offer built-in disaster recovery capabilities, ensuring business continuity in the event of a disruption or outage. With data stored in the cloud, organisations can quickly recover their financial information and resume their operations without significant downtime or data loss.

While the benefits of accelerating cloud adoption in finance are clear, organisations may encounter several challenges along the way. One common challenge is resistance to change. Finance professionals may be hesitant to embrace cloud technology due to concerns about data security, job security, or a lack of familiarity with cloud-based solutions. Overcoming this resistance requires effective communication, training, and a clear understanding of the benefits that cloud adoption brings to finance transformation.

Another challenge is the complexity of migrating existing systems and data to the cloud. Finance departments often have vast amounts of historical data stored in legacy systems, making the migration process complex and time-consuming. It is crucial to carefully plan and execute the migration, ensuring data integrity and minimal disruption to financial operations.

Moreover, organisations need to carefully evaluate and select a reliable cloud provider that meets their specific finance requirements. Factors such as data security, compliance, scalability, and service level agreements (SLAs) should be thoroughly assessed. This process can be daunting, but proper due diligence is essential to ensure a successful cloud adoption journey.

In my experience, to ensure a successful cloud adoption journey in finance transformation, organisations should follow several best practices. Firstly, it is crucial to develop a comprehensive cloud adoption strategy that aligns with the organisation's overall finance transformation goals. This strategy should outline the desired outcomes, timeline, resource allocation, and risk management approach.

Furthermore, organisations should prioritise data security and privacy when selecting a cloud provider. It is essential to choose a provider with robust security measures, data encryption capabilities, and compliance certifications relevant to the finance industry.

Conducting thorough due diligence, including security audits and reference checks, can help organisations make an informed decision.

Another best practice is to start with a pilot project or proof-of-concept before fully migrating to the cloud. This approach allows finance departments to test the waters, identify any potential challenges or limitations, and refine their cloud adoption strategy accordingly. It also helps build confidence among stakeholders and encourages broader adoption across the organisation.

Moreover, organisations should invest in training and upskilling their finance teams to ensure they have the necessary skills and knowledge to leverage cloud technology effectively. This includes training on cloud-based tools, data analytics, and cybersecurity best practices. By empowering finance professionals with the right skills, organisations can maximise the benefits of cloud adoption and drive successful finance transformation.

Several organisations have successfully adopted cloud technology to transform their finance operations. One such example is *PayPal*, which serves more than 300 million users worldwide and processed over 3.74 billion transactions in just the first quarter of 2021.[1]

From the standpoint of a single user, PayPal is just a convenient service for online payments. However, when you look at the whole picture, the company requires astounding operational power to manage the sheer volume of daily transactions while maintaining top security, addressing financial risks, and preventing fraud. All this was made possible after partnering up with Google Cloud in 2018. By accelerating cloud adoption, they were able to centralise their financial data, streamline processes, and gain real-time visibility into their financial performance. This transformation enabled PayPal to make faster, data-driven decisions and achieve significant cost savings.

Another case study is *Bank of America*, a global financial institution. Bank of America recognised the need for enhanced data security and compliance in their finance operations. By partnering with a reputable cloud provider, they were able to strengthen their data protection measures and achieve regulatory compliance more efficiently. By building its cloud, Bank of America saved $2 billion annually (on annual infrastructure savings). It helped reduce the firm's servers to 70,000 from 200,000 and its data centres to 23 from 60.[2] The cloud-based solution also enabled Bank of America to improve its disaster recovery capabilities, ensuring uninterrupted financial services for its customers.

These case studies highlight the transformative power of cloud adoption in finance. By leveraging the cloud, organisations can overcome operational challenges, drive innovation, and achieve their finance transformation goals.

When selecting a *cloud provider* for finance transformation, organisations should consider several key factors. Firstly, data security and privacy should be a top priority. The chosen provider should have robust security measures, data encryption capabilities, and compliance certifications relevant to the finance industry. It is also essential to review the provider's track record and reputation in terms of data breaches or security incidents.

Scalability is another critical consideration. The cloud provider should offer flexible and scalable solutions that can accommodate the organisation's growth and changing needs. This includes the ability to easily add or remove computing resources, as well as scale storage capacity as required.

Additionally, organisations should assess the provider's SLAs to ensure they meet their specific finance requirements. This includes factors such as uptime guarantees, responsiveness to support requests, and data recovery time objectives. It is essential to have a clear understanding of the provider's commitment to service quality and availability.

Lastly, organisations should evaluate the provider's pricing model and cost structure. Cloud services typically operate on a pay-as-you-go or subscription basis. It is crucial to understand the pricing structure and any potential additional costs, such as data transfer fees or storage overages. Conducting a cost analysis and comparing different providers can help organisations make an informed decision that aligns with their budgetary considerations.

Resistance to cloud adoption in finance is not uncommon, but there are strategies to overcome it. One approach is to provide education and training to finance professionals, highlighting the benefits of cloud technology and addressing their concerns. This can be done through workshops, webinars, or one-on-one sessions where employees have the opportunity to ask questions and gain a better understanding of how the cloud can support their work.

Another effective strategy is to involve finance professionals in the decision-making process. By including them in the evaluation and selection of a cloud provider, organisations can ensure that their specific needs and requirements are considered. This involvement can help build trust and ownership among finance professionals, making them more receptive to the idea of cloud adoption.

It is also essential to communicate the long-term vision and benefits of cloud adoption to all stakeholders. This includes finance professionals, senior management, and other departments that may be impacted by the transformation. By demonstrating the positive impact on efficiency, collaboration, and cost savings, organisations can create a compelling case for cloud adoption and gain buy-in from key decision-makers.

As technology continues to evolve, the future of finance transformation lies in accelerated cloud adoption. The cloud offers unparalleled flexibility, scalability, and security, enabling finance departments to adapt to changing business needs and drive innovation. With advancements in artificial intelligence, machine learning, and automation, finance professionals can leverage cloud-based tools to optimise financial processes, improve forecasting accuracy, and enhance decision-making.

Moreover, with the increasing adoption of remote work and the need for virtual collaboration, the cloud becomes even more critical for finance transformation. Cloud-based solutions empower finance professionals to access their data and applications from anywhere, at any time, fostering collaboration and enabling seamless workflows.

In the future, we can expect to see further integration between cloud-based finance systems and other emerging technologies, such as blockchain and the Internet of Things. These technologies have the potential to revolutionise financial processes, enhance transparency, and enable real-time tracking of financial transactions.

Accelerating cloud adoption is essential for successful finance transformation in today's digital age. The cloud offers numerous benefits, including improved scalability, cost savings, enhanced security, and streamlined processes. By embracing cloud technology, finance departments can future-proof their operations, drive innovation, and make data-driven decisions.

However, organisations need to navigate various challenges and considerations when adopting the cloud. Resistance to change, data migration complexities, and selecting the right cloud provider are all factors that must be addressed. By following best practices, involving stakeholders, and prioritising data security, organisations can navigate these challenges and reap the rewards of accelerated cloud adoption.

The future of finance transformation lies in the cloud. Embrace the power of the cloud and unlock the full potential of your finance operations. Start your journey towards successful finance transformation today.

NOTES

1. https://customerthink.com/top-10-companies-using-cloud-and-why/
2. https://www.rishabhsoft.com/blog/cloud-computing-in-banking-and-finance#:~:text=Here's%20some%20insight%20into%20cloud,centers%20to%202023%20from%2060

CHAPTER 14

Quantum Computing

Q UANTUM COMPUTING HAS LONG been a subject of fascination in the realm of science fiction, but in recent years, it has transitioned from the pages of novels to the forefront of technological innovation. With its immense processing power and ability to solve complex problems at an unprecedented speed, quantum computing holds tremendous potential for transforming various industries, including finance.

The field of finance relies heavily on data analysis, risk assessment, and optimisation. Traditional computing systems, while powerful, often struggle to handle the immense complexity and scale of financial calculations. This is where quantum computing comes into play. By harnessing the principles of quantum mechanics, quantum computers can process vast amounts of data simultaneously, enabling them to solve complex financial models and optimise investment strategies with remarkable efficiency.

Saxunova et al. (2020) argue the potential for digital transformation and quantum computing to revolutionise finance is breathtaking. There is no doubt quantum will cause a paradigm shift, but they also acknowledge the greater risk these technologies will cause in terms of security, trust, and fraud issues (Saxunova and le Roux, 2020).

The potential applications of quantum computing in finance are vast. From portfolio optimisation and risk management to fraud detection and algorithmic trading, quantum computing has the ability to revolutionise the way financial institutions operate. The speed and accuracy of quantum algorithms can provide real-time insights, enabling financial professionals to make data-driven decisions with greater precision.

In my experience, the adoption of quantum computing in finance offers numerous benefits. Firstly, it has the potential to significantly enhance computational power, allowing for faster and more accurate data analysis. With quantum computers, financial institutions can process massive datasets in a fraction of the time it would take traditional computers, enabling them to gain valuable insights and make informed decisions in near real-time.

Additionally, quantum computing has the potential to improve risk management in the finance industry. By utilising quantum algorithms, institutions can simulate and analyse complex financial scenarios, enabling them to identify potential risks and develop robust risk mitigation strategies.

DOI: 10.1201/9781003514503-14

Furthermore, quantum computing has the potential to revolutionise the field of cryptography. Quantum computers have the ability to break traditional encryption methods, making them a double-edged sword. While this poses a significant security risk, it also opens up new possibilities for developing quantum-resistant encryption algorithms, ensuring the security of financial transactions in the future.

Although quantum computing is still in its early stages, the finance industry has already begun exploring its potential applications. For instance, *JP Morgan Chase* has been actively researching quantum algorithms for portfolio optimisation, aiming to improve investment strategies and maximise returns for their clients.

Another notable example is *Goldman Sachs*, which has partnered with quantum computing companies to explore the use of quantum algorithms in option pricing and risk analysis. By leveraging quantum computing, Goldman Sachs aims to enhance its trading strategies and gain a competitive edge in the market.

Furthermore, quantum computing is being utilised in the field of credit scoring. Traditional credit scoring models often rely on limited data points, leading to inaccurate assessments of creditworthiness. Quantum computing can enable financial institutions to process vast amounts of data, allowing for more accurate credit scoring and risk assessment, ultimately leading to fairer lending practices.

Quantum computing has emerged as a ground-breaking technology with the potential to revolutionise various industries. One sector where it has shown immense promise is finance, particularly in the realm of *risk management*. Traditional risk management models often struggle to accurately assess and mitigate complex financial risks due to their limited computational power. However, quantum computers possess the ability to process vast amounts of data simultaneously, allowing for more sophisticated risk analysis and management strategies.

With quantum computing, financial institutions can enhance their risk management capabilities by leveraging the technology's ability to perform complex calculations at an unprecedented speed. By harnessing the power of quantum algorithms, these institutions can better predict market fluctuations, analyse intricate financial models, and identify potential risks that were previously undetectable. This quantum-powered risk management approach enables more accurate assessments of portfolio risk, leading to better-informed investment decisions and ultimately reducing overall exposure to potential financial losses.

Furthermore, quantum computing enables the development of more robust risk models that can handle a wider range of variables and scenarios. Traditional models often rely on simplifications and assumptions that do not fully capture the complexity and interdependencies within financial markets. However, quantum algorithms can incorporate a multitude of factors, including real-time market data, economic indicators, and historical trends, allowing for more accurate risk assessments. This enhanced risk modelling capability is particularly crucial in today's rapidly evolving financial landscape, where unexpected events can have significant repercussions on global markets.

The integration of quantum computing into risk management practices has the potential to transform the way financial institutions identify, assess, and manage risks. By leveraging the immense computational power of quantum computers, these institutions can enhance their risk management capabilities, leading to more informed decision-making and improved overall financial stability.

Algorithmic trading has become a cornerstone of modern finance, enabling institutions to execute trades with speed and precision. However, as financial markets become increasingly complex and volatile, traditional computing systems are often unable to keep up with the demands of algorithmic trading. This is where quantum computing steps in, offering the potential to revolutionise this crucial aspect of finance.

Quantum computers possess the ability to perform complex calculations simultaneously, providing a substantial advantage in the world of algorithmic trading. The sheer computational power of quantum algorithms allows for the analysis of vast amounts of data, enabling traders to identify patterns and trends that were previously unattainable. This enhanced data analysis capability empowers algorithmic traders to make more informed decisions, optimise their strategies, and increase profitability.

Moreover, quantum computing can significantly speed up the execution of trades. In today's fast-paced financial markets, even a fraction of a second can make a substantial difference in trade outcomes. Quantum computers excel at solving optimisation problems, such as determining the optimal order in which to execute trades to minimise transaction costs or maximise profits. By leveraging quantum algorithms, traders can accelerate their trading strategies, gaining a competitive edge in the market.

However, it is important to note that the full integration of quantum computing in algorithmic trading is still in its infancy. The technology is complex and requires further development and refinement before it can be widely adopted. Additionally, there are challenges related to the stability and reliability of quantum systems, as well as the need for specialised expertise in quantum programming. Nonetheless, the potential of quantum computing in revolutionising algorithmic trading is undeniable, and financial institutions are closely monitoring its progress and investing in research and development.

Fraud detection is a critical concern for financial institutions, as the costs associated with fraudulent activities can be substantial. Traditional methods of fraud detection often rely on rule-based systems that struggle to keep up with the ever-evolving tactics employed by fraudsters. Quantum computing, with its ability to process vast amounts of data simultaneously, offers a potential solution to this ongoing challenge.

By harnessing the power of quantum algorithms, financial institutions can significantly enhance their fraud detection capabilities. Quantum computers can analyse large datasets in real-time, identifying patterns and anomalies that may indicate fraudulent activities. The speed and efficiency of quantum computing enable institutions to detect and respond to potential fraud attempts swiftly, minimising financial losses and protecting the integrity of their operations.

Furthermore, quantum computing can help financial institutions improve their fraud prevention strategies. Traditional systems often rely on a reactive approach, detecting fraud after it has occurred. Quantum algorithms, on the other hand, can be used to develop proactive fraud prevention models that can identify potential vulnerabilities in real-time. By continuously monitoring and analysing data, these models can detect subtle indicators of fraudulent behaviour, allowing institutions to implement preventive measures before any harm is done.

However, it is essential to acknowledge that the implementation of quantum computing in fraud detection is still in its early stages. There are challenges to overcome, such as the integration of quantum systems with existing infrastructure and the development of robust quantum algorithms specifically tailored for fraud detection. Nonetheless, the potential benefits of quantum computing in combating financial fraud are vast, and financial institutions are actively exploring ways to leverage this technology.

Portfolio optimisation is a key aspect of finance, aiming to maximise returns while minimising risk. Traditionally, portfolio optimisation has relied on mathematical models that make various assumptions and simplifications, which may not accurately capture the complexities of financial markets. Quantum computing offers a new paradigm for portfolio optimisation, enabling more sophisticated and accurate strategies.

Quantum computers can process vast amounts of data and simultaneously consider multiple variables, allowing for a more comprehensive analysis of portfolio optimisation. By leveraging quantum algorithms, financial institutions can identify optimal investment strategies that take into account a broader range of factors, including market trends, economic indicators, and real-time data. This enhanced optimisation capability enables institutions to construct portfolios that are better aligned with their investment objectives and risk tolerance.

Moreover, quantum computing can address one of the major challenges in portfolio optimisation – the curse of dimensionality. As the number of assets in a portfolio increases, the computational complexity of traditional optimisation methods grows exponentially.

Quantum algorithms offer the potential to overcome this challenge by efficiently exploring the vast solution space, enabling the identification of optimal portfolios even for large-scale investment scenarios.

However, it is vital to acknowledge that the adoption of quantum computing in portfolio optimisation is still in its early stages. There are practical considerations, such as the need for hardware advancements and the development of specialised quantum algorithms for portfolio optimisation. Additionally, the integration of quantum systems with existing financial infrastructure poses challenges that need to be addressed. Nonetheless, the potential of quantum computing in revolutionising portfolio optimisation is significant, and financial institutions are actively exploring its applications.

While the potential of quantum computing in finance is immense, there are several challenges and limitations that need to be addressed. Firstly, quantum computers are still in the early stages of development, and building reliable and scalable quantum systems remains a significant challenge. The technology is highly complex and requires specialised experts to design and operate quantum computers effectively.

Additionally, quantum computing is highly sensitive to noise and errors. Quantum bits, or qubits, are fragile and prone to disturbances from their environment. As a result, maintaining the stability of qubits and minimising errors is crucial for accurate calculations. Overcoming these challenges will require advancements in quantum error correction and fault-tolerant quantum computing.

Another limitation is the high cost of quantum computing. Building and maintaining quantum computers is an expensive endeavour. The current cost of quantum hardware and infrastructure is prohibitive for many financial institutions, making it difficult to integrate quantum computing into existing systems and processes. As the technology progresses and becomes more accessible, these costs are expected to decrease, making quantum computing more feasible for widespread adoption.

Despite the challenges, financial institutions can take several *strategic approaches* to implement quantum computing in their transformation journey. Firstly, collaboration and partnerships with quantum computing companies can provide access to the expertise and resources necessary for building quantum algorithms and models specific to the finance industry.

Additionally, financial institutions can invest in research and development to explore the potential applications of quantum computing in their specific domains. By dedicating resources to understand the nuances of quantum computing and its impact on finance, organisations can position themselves at the forefront of innovation and gain a competitive advantage.

Moreover, financial institutions can leverage cloud-based quantum computing platforms. Several tech giants, such as IBM and Microsoft, offer cloud-based access to quantum computers, allowing organisations to experiment and develop quantum algorithms without the need for significant upfront investment in hardware.

While quantum computing is still in its infancy, there have been notable successes in utilising quantum algorithms for finance applications. For example, *HSBC* partnered with a quantum computing company to test the use of quantum annealing for portfolio optimisation. The results showed a significant reduction in the time required for optimisation, demonstrating the potential of quantum computing in improving investment strategies.

Another case study involves the use of quantum computing for derivative pricing. *Banco Santander* collaborated with a quantum computing company to develop a quantum algorithm for pricing options. The algorithm demonstrated superior performance compared to traditional methods, highlighting the potential of quantum computing in enhancing pricing models in the finance industry.

The integration of quantum computing in finance transformation holds immense potential for reshaping the industry. The technologies discussed earlier – risk management, algorithmic trading, fraud detection, and portfolio optimisation – are just a glimpse into the possibilities that quantum computing offers. As quantum computing continues to evolve and mature, its impact on finance is expected to grow exponentially.

In the future, we can anticipate quantum computing to transform financial operations across various fronts. The technology's ability to process vast amounts of data simultaneously will enable institutions to gain deeper insights into market trends, customer behaviour, and risk factors. This enhanced understanding will drive more informed decision-making and enable financial institutions to provide more tailored and personalised services to their clients.

Moreover, as quantum algorithms become more sophisticated, we can expect the development of new financial products and services that leverage the capabilities of quantum computing. These innovations may include advanced derivatives pricing models, optimised asset allocation strategies, and more accurate credit risk assessments. The integration of quantum computing in finance has the potential to unlock new avenues for growth and profitability.

However, the realisation of quantum computing's full potential in finance transformation requires collaboration and investment across academia, industry, and government. Further research and development are needed to address the challenges associated with quantum computing, such as hardware stability, algorithmic refinement, and integration with existing financial systems. Additionally, specialised expertise in quantum programming and quantum finance will be essential to harness the power of this technology effectively.

Quantum computing is poised to revolutionise finance transformation. From risk management to algorithmic trading, fraud detection, and portfolio optimisation, the potential applications of quantum computing in finance are vast. As technology continues to evolve, financial institutions must embrace the opportunities it presents and invest in research and development to stay at the forefront of this exciting transformation.

As quantum computing continues to evolve, its impact on the finance industry is poised to be transformative. With its ability to process vast amounts of data and solve complex problems, quantum computing has the potential to revolutionise financial services, from risk management and asset pricing to fraud detection and algorithmic trading.

However, it is important to note that the integration of quantum computing into the finance industry will be a gradual process. As the technology matures and becomes more accessible, financial institutions will need to adapt their systems, infrastructure, and workforce to harness its full potential.

For finance professionals looking to gain a deeper understanding of quantum computing, there are several resources available. Online platforms such as *Quantum Computing for Finance* provide courses and tutorials specifically tailored to the finance industry, covering topics such as quantum algorithms, quantum risk analysis, and quantum cryptography.

Additionally, *research papers and publications* by leading quantum computing researchers and institutions can provide valuable insights into the current state and future prospects of quantum computing in finance.

As we venture further into the digital age, the finance industry must embrace emerging technologies to stay ahead of the curve. Quantum computing represents a paradigm shift in computational power, offering unprecedented capabilities that can transform the way financial institutions operate.

While there are challenges and limitations to overcome, the potential benefits of harnessing quantum computing in finance are immense. From improving data analysis and risk management to enhancing decision-making and developing quantum-resistant encryption, quantum computing has the potential to revolutionise the finance industry.

As the field continues to evolve and quantum computers become more accessible, financial professionals must stay informed and be prepared to adapt. By embracing the quantum revolution, financial institutions can position themselves at the forefront of innovation and drive transformative change in the finance industry.

CHAPTER 15

Blockchain Technology

B LOCKCHAIN TECHNOLOGY HAS EMERGED as a revolutionary force in the world of finance. It has the potential to transform traditional financial systems by introducing transparency, efficiency, and security. Chief Finance Officer (CFO) should consider the power of blockchain and its role in finance transformation.

Blockchain is a decentralised, distributed ledger that allows for the secure and transparent recording of transactions. It operates on a network of computers, known as nodes, where each node has a copy of the entire blockchain. Every transaction is recorded in a block, which is then added to the chain of previous blocks, forming a chronological and immutable record.

Finance transformation is a vital process of reimagining and restructuring financial operations to enhance efficiency, effectiveness, and value creation. CFOs must ensure the adoption of new technologies, methodologies, and strategies to streamline financial processes and drive growth.

Traditionally, financial systems have relied on centralised authorities, such as banks and intermediaries, to facilitate transactions and maintain records. However, these systems are often slow, prone to errors, and lack transparency. Finance transformation aims to address these limitations and create a more agile and resilient financial ecosystem.

Blockchain technology plays a pivotal role in finance transformation by providing a decentralised and secure platform for financial transactions. It eliminates the need for intermediaries, reduces transaction costs, and enhances transaction speed. Additionally, blockchain enables the creation of smart contracts, which are self-executing contracts with predefined rules and conditions.

Wang (2023) argues blockchain has huge potential in the finance sector and that the most effective applications they have seen of emerging technologies in FinTechs involve blockchain (specifically cryptocurrencies), robo-advisors, online payment services, and peer-to-peer lending platforms (Wang, 2023). Gao (2024) also argues blockchain will have a significant impact thanks to P2P lending and cryptocurrencies (Gao, 2024).

By leveraging blockchain, financial institutions can streamline processes, improve record-keeping, and enhance security. It allows for real-time auditing, eliminates the risk

DOI: 10.1201/9781003514503-15

of data manipulation, and ensures the integrity of financial transactions. Furthermore, blockchain enables seamless cross-border transactions and enhances financial inclusion by providing access to financial services for the unbanked population.

The adoption of blockchain technology in finance offers numerous benefits. First and foremost, it enhances transparency and trust by providing a verifiable and auditable record of transactions. This transparency reduces the risk of fraud and improves regulatory compliance.

Secondly, blockchain improves efficiency by automating processes and reducing the need for manual reconciliation. It eliminates the need for intermediaries, such as clearing-houses and custodians, thereby reducing costs and settlement times.

Furthermore, blockchain enhances security by utilising advanced cryptographic algorithms and consensus mechanisms. The decentralised nature of blockchain makes it resistant to hacking and data tampering. This increased security is particularly crucial in the finance industry, where the confidentiality and integrity of financial data are paramount.

Blockchain technology has already made significant strides in the finance industry. One notable example is the use of blockchain for *cross-border payments*. Traditional cross-border transactions are often slow, costly, and subject to intermediaries. However, blockchain-based platforms enable near-instantaneous and low-cost cross-border transfers, eliminating the need for intermediaries and reducing transaction fees.

Another application of blockchain in finance is *trade finance*. Blockchain can streamline the complex process of trade finance by digitising and automating documentation, such as letters of credit and bills of lading. This reduces paperwork, eliminates the risk of fraud, and accelerates the settlement process.

Additionally, blockchain has the potential to revolutionise the *lending and borrowing processes*. By leveraging blockchain, lenders can securely verify borrowers' identities, assess their creditworthiness, and automate loan disbursements. This reduces the risk of default and improves access to credit for individuals and businesses.

While blockchain technology holds immense promise, it also faces several challenges and limitations. One of the main challenges is scalability. As the number of transactions on the blockchain increases, the network can become congested, leading to slower transaction speeds. Additionally, the storage requirements for maintaining a copy of the entire blockchain can be significant.

Another challenge is regulatory compliance. The decentralised nature of blockchain makes it difficult to regulate and enforce existing financial regulations. Governments and regulatory bodies are still grappling with how to strike a balance between innovation and security in the face of blockchain technology.

Furthermore, there is a need for interoperability and standardisation in blockchain networks. Different blockchain platforms and protocols may not be compatible with each other, hindering seamless integration and collaboration.

In my experience, implementing blockchain in finance transformation requires careful planning and consideration. Here are some key steps to follow:

Identify Use Cases: Determine the specific areas in finance where blockchain can add value, such as cross-border payments, trade finance, or supply chain finance.

Evaluate Existing Systems: Assess the current financial infrastructure and identify pain points and inefficiencies that can be addressed through blockchain.

Choose the Right Blockchain Platform: Select a blockchain platform that aligns with the organisation's requirements in terms of scalability, security, and functionality.

Collaborate with Stakeholders: Engage with relevant stakeholders, including regulators, financial institutions, and technology providers, to ensure a smooth implementation and address any concerns.

Develop a Proof of Concept: Create a proof of concept to demonstrate the feasibility and benefits of blockchain in the identified use case. This will help garner support and investment for the full-scale implementation.

Pilot Implementation: Start with a small-scale pilot implementation to test the viability and scalability of the blockchain solution. This will allow for iterative improvements and refinements before wider adoption.

Successful integration of blockchain in finance transformation requires attention to several key considerations:

Security: Implement robust security measures, such as encryption and multi-factor authentication, to protect sensitive financial data.

Compliance: Ensure compliance with relevant regulations and industry standards, such as KYC (Know Your Customer) and AML (Anti-Money Laundering) requirements.

Governance: Establish clear governance structures and frameworks to ensure transparency, accountability, and regulatory compliance.

Collaboration: Foster collaboration and partnerships with other organisations in the finance industry to drive innovation and standardisation.

Education and Training: Invest in educating and training employees on blockchain technology to facilitate adoption and maximise its potential.

The future of blockchain in finance transformation looks promising. Here are some key trends and predictions:

Central Bank Digital Currencies: Central banks around the world are exploring the idea of issuing digital currencies based on blockchain technology. This could revolutionise the concept of money and reshape the financial landscape.

Tokenisation of Assets: Blockchain enables the digitisation and fractional ownership of assets, such as real estate, stocks, and commodities. This opens up new investment opportunities and enhances liquidity.

Integration with Emerging Technologies: Blockchain is likely to be integrated with other emerging technologies, such as artificial intelligence and the Internet of Things (IoT). This convergence will enable new business models and enhance the value proposition of blockchain.

Industry-Wide Collaboration: As blockchain matures, we can expect increased collaboration among financial institutions, technology providers, and regulators to develop industry-wide standards and frameworks.

Digital currencies have emerged as a revolutionary force in the financial industry, offering new possibilities for transactions and investments. In recent years, the concept of tokenisation has gained traction, further expanding the potential of digital currencies. Tokenisation is an important process that converts real-world assets into digital tokens, representing ownership or access rights. These tokens are then recorded on a blockchain, ensuring transparency, security, and immutability.

Tokenisation is a fundamental concept in the world of digital currencies. It involves the representation of real-world assets, such as real estate, art, or even stocks, as digital tokens on a blockchain. These tokens can be easily traded, transferred, and divided into fractional shares, providing liquidity and accessibility to a wide range of investors. The underlying technology behind tokenisation is blockchain, a decentralised ledger that ensures the integrity and security of transactions.

Digital currencies, on the other hand, are virtual or digital representations of value that are used as a medium of exchange. Unlike traditional fiat currencies issued by central banks, digital currencies are typically decentralised and operate on a peer-to-peer network. They offer advantages such as faster and cheaper transactions, increased security, and greater financial inclusion. With the advent of tokenisation, digital currencies have become even more versatile and powerful.

Blockchain technology plays a crucial role in enabling tokenisation. It serves as a decentralised and transparent ledger that records and verifies transactions. By using blockchain, tokenisation ensures the integrity and immutability of digital assets, making them secure and resistant to fraud. Blockchain also enables the efficient transfer of tokens, eliminating the need for intermediaries, reducing costs, and increasing transaction speed.

Blockchain technology enables the creation of smart contracts, which are self-executing contracts with the terms of the agreement directly written into lines of code. Smart contracts facilitate the automation of transactions, ensuring that all parties involved fulfil their obligations. This automation streamlines processes, reduces the risk of fraud, and enhances trust among participants.

Tokenisation brings several advantages to the financial industry. Firstly, it provides increased liquidity and accessibility to assets that were traditionally illiquid or difficult to divide. For example, tokenisation allows individuals to own fractional shares of a high-value asset, such as a luxury property or a rare piece of art. This fractional ownership enables a wider range of investors to participate and diversify their portfolios.

Secondly, tokenisation reduces the barriers to entry for investors. Traditionally, investing in certain assets required significant capital and was limited to institutional investors or high-net-worth individuals. With tokenisation, anyone can invest in a wide range of assets, regardless of their financial status. This democratisation of investment opportunities empowers individuals and promotes financial inclusion.

Thirdly, tokenisation enhances transparency and security. By recording transactions on a blockchain, tokenisation ensures transparency and immutability, reducing the risk of fraud or manipulation. Additionally, the use of blockchain technology provides enhanced security measures, protecting digital assets from cyber threats.

Tokenisation has already found numerous use cases in the financial industry. One prominent example is the tokenisation of *real estate*. By converting real estate assets into digital tokens, individuals can invest in properties with smaller amounts of capital. This opens up opportunities for real estate investment to a wider audience, fostering financial inclusion and diversification.

Another use case is the tokenisation of *stocks and other securities*. Tokenising stocks allows for easy and efficient trading, eliminating the need for intermediaries and reducing transaction costs. It also enables fractional ownership, making it possible for individuals to invest in high-value stocks without having to purchase a whole share.

Tokenisation is also being explored in the *lending and borrowing* space. By tokenising debt instruments, such as loans or bonds, lenders can easily transfer ownership and trade these assets. This enhances liquidity in the market and provides new investment opportunities for individuals.

Tokenisation has the potential to *disrupt traditional finance systems* in several ways. Firstly, it challenges the role of intermediaries, such as banks or brokers, by enabling peer-to-peer transactions. This disintermediation reduces costs and increases efficiency in financial transactions.

Secondly, tokenisation introduces programmable assets through the use of smart contracts. These programmable assets can automatically execute predefined conditions, such as dividend payments or voting rights. This automation eliminates the need for manual intervention, reducing administrative overhead and streamlining processes.

Lastly, tokenisation enables new forms of fundraising, such as initial coin offerings (ICOs) or security token offerings (STOs). These fundraising methods allow companies to raise capital by offering digital tokens representing ownership or future rights. This alternative to traditional fundraising methods has the potential to democratise access to capital and foster innovation.

While tokenisation brings significant advantages, it also poses challenges and risks. One challenge is regulatory compliance. As tokenisation blurs the boundaries between traditional assets and digital tokens, regulatory frameworks need to adapt to ensure investor protection, prevent fraud, and maintain market integrity. Clear guidelines and regulations are necessary to provide certainty and foster trust in tokenised assets.

Another challenge is the interoperability of different blockchain platforms. Currently, there are multiple blockchain networks, each with its own protocols and standards. Interoperability between these networks is crucial for the seamless transfer of assets and the integration of various services. Efforts are being made to develop interoperability solutions, but further advancements are needed.

Additionally, tokenisation faces risks related to cybersecurity and privacy. As digital assets become more valuable, they become attractive targets for hackers. Ensuring robust security measures and protecting user data are critical for the widespread adoption of tokenisation.

The future of tokenisation looks promising, with several trends and opportunities on the horizon. One trend is the increasing tokenisation of intangible assets, such as intellectual property or patents. Tokenising these assets can unlock their value and enable creators to monetise their work more easily.

Another trend is the integration of tokenisation with IoT devices. By combining tokenisation with IoT, physical assets can be represented as digital tokens and seamlessly transferred or traded. This opens up new possibilities for asset ownership and usage.

Furthermore, decentralised finance (DeFi) is an emerging field that leverages tokenisation and blockchain technology to create innovative financial products and services. DeFi aims to provide an open and inclusive financial infrastructure, enabling individuals to access financial services without relying on traditional intermediaries.

Tokenisation is revolutionising the financial industry by unlocking the potential of digital currencies. It enables the representation of real-world assets as digital tokens on a blockchain, offering increased liquidity, accessibility, transparency, and security. While there are challenges and risks associated with tokenisation, the future holds exciting opportunities for further innovation and disruption in the financial sector. As digital currencies and tokenisation continue to evolve, it is crucial for regulators, industry players, and investors to embrace this transformative technology and adapt to the changing landscape.

Blockchain technology has the potential to revolutionise finance transformation by introducing transparency, efficiency, and security. By leveraging blockchain, financial institutions can streamline processes, reduce costs, and enhance trust. However, the adoption of blockchain in finance transformation comes with its own set of challenges and limitations. It requires careful planning, collaboration, and consideration of key factors for successful integration. As we look to the future, blockchain is poised to reshape the financial landscape and unlock new possibilities for innovation and growth. It is imperative for organisations in the finance industry to embrace the potential of blockchain and stay ahead of the curve.

CHAPTER 16

Web 3

T HE RAPID ADVANCEMENT OF technology has brought about significant changes in various industries, including finance. As we enter the digital era, the concept of Web 3 has emerged, revolutionising the way finance operates. Web 3, also known as the decentralised web, is a new paradigm that aims to empower individuals and create a more transparent, secure, and efficient financial system. CFOs should consider Web 3 and its implications for finance transformation.

Lekhi (2023) argues Web 3.0 introduces new technologies such as blockchain, decentralized apps, and smart contracts that enable secure, transparent, and decentralised financial transactions. Lekhi contends this will reduce the need for intermediaries, decrease transaction costs, and enhance the speed and efficiency of financial processes. Web 3.0 allows for the creation of new financial products and services that were previously impossible, such as decentralised exchanges, non-fungible token (NFT), peer-to-peer lending platforms, and decentralised autonomous organisations. These innovations could democratize access to financial services and provide new opportunities for individuals and businesses alike (Lekhi, 2023).

Finance transformation will mean a shift to enhance efficiency, reduce costs, and improve decision-making. Traditionally, finance has been centralised, with financial institutions acting as intermediaries in transactions. However, with the advent of Web 3, the focus is shifting towards decentralisation, allowing individuals to directly interact and transact with each other without the need for intermediaries. This shift has the potential to disrupt the traditional finance industry and open up new opportunities for innovation and growth.

The digital era has transformed the way we live, work, and interact. The rise of the internet and digital technologies has enabled the digitisation of financial services, making it easier for individuals and businesses to access financial products and services. However, the centralised nature of the current financial system has its limitations, including high transaction fees, limited accessibility, and a lack of transparency. Web 3 aims to address these limitations by leveraging blockchain technology and decentralised platforms to create a more inclusive and efficient financial ecosystem.

Web 3 introduces several key features that are transforming finance. One of the key features is decentralisation, which eliminates the need for intermediaries and enables

DOI: 10.1201/9781003514503-16

peer-to-peer transactions. This reduces costs, increases transparency, and improves the speed of transactions. Another key feature is smart contracts, which are self-executing contracts with the terms of the agreement directly written into code. Smart contracts automate and streamline processes, reducing the need for manual intervention and reducing the risk of errors. Additionally, Web 3 introduces the concept of digital identity, which allows individuals to have control over their personal data and securely verify their identity online. These features of Web 3 are revolutionising finance by enabling a more efficient, secure, and inclusive financial ecosystem.

The adoption of Web 3 in finance offers numerous benefits and advantages. Firstly, it provides individuals with greater control over their financial transactions and data. With Web 3, individuals can securely store and manage their financial assets, eliminating the need for third-party custodians. Secondly, Web 3 reduces transaction costs by eliminating intermediaries and automating processes. This can lead to significant cost savings for both individuals and businesses. Thirdly, Web 3 enhances transparency and trust in financial transactions. The use of blockchain technology ensures that transactions are recorded and verified in a transparent and immutable manner, reducing the risk of fraud and corruption. Lastly, Web 3 promotes financial inclusion by providing access to financial services for individuals who are currently underserved by the traditional financial system. This has the potential to improve the livelihoods of millions of people around the world.

While Web 3 offers numerous benefits, there are also challenges and considerations that need to be addressed in implementing it for finance transformation. One of the main challenges is regulatory compliance. As Web 3 operates on decentralised platforms, it may be difficult to comply with existing financial regulations. This requires collaboration between regulators and industry stakeholders to create a regulatory framework that ensures consumer protection while fostering innovation. Another challenge is scalability. The current blockchain infrastructure may not be able to handle the scale of financial transactions required for mainstream adoption. This calls for the development of scalable solutions that can handle high transaction volumes without compromising security and efficiency. Additionally, there is a need for education and awareness to ensure that individuals and businesses understand the potential of Web 3 and are able to effectively utilise its features.

Several organisations have already embraced Web 3 to transform their finance operations. One such example is *MakerDAO*, a decentralised autonomous organisation that provides a stablecoin called DAI. DAI is pegged to the US dollar and backed by crypto assets, allowing individuals to transact and store value without the need for traditional banking services. Another example is *Aave*, a decentralised lending platform that allows users to borrow and lend crypto assets. Aave uses smart contracts to automate the lending process, eliminating the need for intermediaries and reducing costs. These case studies demonstrate the potential of Web 3 in transforming finance and creating new opportunities for innovation and growth.

In my experience, implementing Web 3 in finance transformation requires a *strategic approach*. Organisations need to assess their existing financial systems and processes and identify areas where Web 3 can bring the most value. This may involve conducting a thorough analysis of the organisation's financial operations, identifying pain points, and exploring how Web 3 can address these challenges. Organisations also need to develop a clear roadmap for implementation, considering factors such as regulatory compliance, scalability, and security. Collaborating with industry partners and engaging with regulators can also help organisations navigate the complexities of implementing Web 3 in finance transformation.

The future of Web 3 in the finance industry looks promising. As technology continues to advance, we can expect to see further innovation and development in the field of decentralised finance. One of the key trends that we can expect is the integration of Web 3 with other emerging technologies, such as artificial intelligence and the Internet of Things. This integration has the potential to create new opportunities for automation, data analysis, and personalised financial services. We can also expect to see the emergence of new financial products and services that leverage the power of Web 3, such as decentralised exchanges, tokenised assets, and decentralised lending platforms. Overall, the future of Web 3 in the finance industry is exciting, and organisations that embrace this technology have the potential to gain a competitive advantage.

Web 3 is revolutionising finance transformation in the digital era. By leveraging blockchain technology and decentralised platforms, Web 3 is empowering individuals, reducing costs, improving transparency, and promoting financial inclusion. While there are challenges and considerations in implementing Web 3, organisations that embrace this technology have the potential to transform their finance operations and gain a competitive advantage. As we move forward, it is essential for organisations to stay informed about the latest developments in Web 3 and explore how it can be integrated into their financial systems and processes. By embracing the potential of Web 3, organisations can pave the way for a more efficient, secure, and inclusive financial ecosystem.

CHAPTER 17

Internet of Things (IoT)

T HE INTERNET OF THINGS (IoT) has revolutionised various industries, and finance is no
exception. IoT refers to the network of interconnected devices that communicate and
exchange data with each other through the internet. In the finance sector, this technology
has the potential to drive transformation and innovation, leading to enhanced efficiency,
improved customer experience, and greater profitability.

IoT devices in finance include sensors, wearables, mobile devices, and even smart
appliances. These devices collect and transmit data in real-time, allowing financial
institutions to gain valuable insights and make data-driven decisions. For example, IoT
devices can monitor market trends, track customer behaviour, and provide personalised
financial advice.

Aznag et al. (2022) IoT has huge potential, with connected payment products through
wearables, connected cars, and smart home devices among the most important categories
of IoT applications in banking and financial technology. Issues related to digital payments
and security are the current area of challenges for the Internet of Things, and Visa has a
unique ability to lead IoT payment innovations (Aznag and Tahanout, 2022).

The rapid advancement of technology has undoubtedly transformed various industries,
and the finance sector is no exception. It is important for CFOs to consider the transformative power of the IoT in finance. From payment systems and transactions to risk management and customer experience, IoT can make a significant impact.

The implementation of IoT in the finance industry has the power to transform various
aspects of financial services, including banking, insurance, and investment management.
By leveraging IoT technologies, financial institutions can automate processes, streamline operations, and offer innovative products and services to meet the evolving needs of
customers.

One of the key areas where IoT can drive finance transformation is *risk management*.
IoT devices can collect and analyse vast amounts of data, enabling financial institutions to
assess risks in real time and make proactive decisions. For instance, IoT sensors can monitor environmental conditions and alert insurers about potential risks, such as flooding or
fire, allowing them to take preventive measures and minimise losses.

DOI: 10.1201/9781003514503-17

Moreover, IoT can also revolutionise *payment systems*. With the introduction of IoT-enabled payment devices, transactions can become more seamless and secure. For instance, smartwatches equipped with IoT technology can facilitate contactless payments, eliminating the need for physical cards or cash. This not only enhances convenience for customers but also reduces the risk of fraud.

The IoT has become a game-changer in the world of finance, bringing about numerous benefits and opportunities. By enabling seamless connectivity and data sharing between devices, IoT has transformed traditional finance processes and paved the way for innovative solutions. From streamlining payment systems to enhancing risk management, IoT has proven to be a powerful tool in driving financial transformation.

In my experience, the implementation of IoT in the finance industry offers numerous benefits for financial institutions, customers, and the overall economy. Firstly, IoT enables financial institutions to collect real-time data, allowing them to gain valuable insights into customer behaviour, market trends, and operational efficiency. This data-driven approach enables institutions to make informed decisions, improve customer experience, and enhance profitability.

Secondly, IoT can improve operational efficiency by automating manual processes and reducing human error. For example, IoT sensors can monitor inventory levels, automatically placing orders when stock is low, thus ensuring that businesses always have the necessary supplies. This automation not only saves time but also reduces costs and improves overall productivity.

Thirdly, IoT can enhance customer experience by providing personalised and innovative financial services. For instance, banks can leverage IoT devices to offer personalised financial advice based on customers' spending patterns and financial goals. This level of personalisation not only increases customer satisfaction but also builds long-term relationships with clients.

Several financial institutions have already embraced IoT and reaped the benefits of its implementation. One such example is the use of IoT in asset management. Asset management companies can leverage IoT sensors to monitor the condition and location of assets, such as vehicles or machinery, in real-time. This enables them to optimise asset utilisation, reduce maintenance costs, and improve overall operational efficiency.

Another successful application of IoT in finance is fraud detection and prevention. By analysing patterns and anomalies in data collected from various IoT devices, financial institutions can detect fraudulent activities in real-time and take immediate action. For example, if an IoT device detects unauthorised access to a customer's account, it can trigger an alert and suspend the transaction, preventing potential financial losses.

Furthermore, IoT has also been utilised in insurance to provide usage-based policies. By installing IoT devices in vehicles, insurance companies can monitor driving behaviour, such as speed, acceleration, and braking patterns. This data is then used to calculate insurance premiums, rewarding safe drivers with lower rates. This not only encourages responsible driving but also provides fairer insurance pricing for customers.

The adoption of IoT in *financial services* has led to the emergence of various use cases that demonstrate the potential of this technology. One such use case is the implementation of IoT in customer service. Financial institutions can leverage IoT devices, such as chatbots or virtual assistants, to provide personalised and interactive customer support. These devices can answer customer queries, provide financial advice, and even assist in completing transactions.

Another use case is the integration of IoT with asset tracking and management. Financial institutions can utilise IoT sensors to track the location and condition of assets, such as high-value equipment or vehicles. This enables institutions to optimise asset utilisation, reduce theft or loss, and improve overall asset management efficiency.

Additionally, IoT can be used in risk management to monitor and mitigate potential risks. For example, financial institutions can install IoT sensors in buildings to monitor environmental conditions, such as temperature or humidity. This allows institutions to detect and prevent risks, such as fire or water damage, and take immediate action to minimise losses.

We can expect increased adoption of IoT in regulatory compliance. Financial institutions can leverage IoT devices to monitor and report compliance-related data, ensuring adherence to regulations and minimising the risk of penalties.

IoT-powered predictive analytics will play a significant role in enhancing financial decision-making. By analysing real-time data from IoT devices, financial institutions can make accurate predictions regarding market trends, customer behaviour, and risk factors. This foresight enables businesses to make informed decisions and stay ahead in the competitive finance landscape.

In the realm of *payment systems and transactions*, the IoT has introduced a new level of convenience and security. Take the case of smart payment devices, such as contactless cards and mobile wallets. These devices leverage IoT technology to enable quick and hassle-free transactions. By connecting to the internet and utilising near-field communication technology, users can simply tap their devices to make payments effortlessly.

Not only does IoT enhance the speed and ease of payments, but it also strengthens security measures. With IoT-enabled devices, transactions are encrypted and authenticated in real-time, significantly reducing the risk of fraud. For instance, smart ATMs equipped with IoT sensors can detect suspicious activities and promptly alert the respective authorities. This level of security instils trust in customers and ensures the integrity of financial transactions.

Risk management and fraud prevention are critical aspects of the finance industry, and IoT has emerged as a valuable tool in combating these challenges. Through IoT-enabled devices, financial institutions can gather vast amounts of data in real-time, allowing them to identify and mitigate risks promptly. For example, in the insurance sector, IoT sensors installed in vehicles can collect information on driving behaviour, enabling insurers to offer personalised and fair premiums based on actual risk factors.

Furthermore, IoT technology enables continuous monitoring of financial transactions, detecting anomalies and potential fraud. By analysing patterns and deviations in data, financial institutions can proactively identify fraudulent activities and take appropriate measures. This not only saves businesses from significant financial losses but also safeguards the interests of their customers.

Customer experience and engagement are vital for any industry, and the finance sector is no different. The IoT has transformed the way financial institutions interact with their customers, enhancing their experience and fostering long-term engagement. For example, IoT-powered chatbots and virtual assistants enable customers to access personalised financial advice and assistance around the clock.

Moreover, IoT devices, such as smartwatches and fitness trackers, have paved the way for innovative banking experiences. Customers can now track their financial health in real-time, receive personalised recommendations, and set financial goals. This level of engagement not only strengthens the relationship between financial institutions and their customers but also empowers individuals to make informed financial decisions.

Efficient asset management and tracking are crucial for the finance industry, and the IoT has revolutionised these processes. IoT-enabled sensors and trackers provide real-time data on the location, condition, and performance of assets, ensuring optimal utilisation and maintenance. For instance, in the banking sector, IoT technology enables real-time monitoring of ATMs, ensuring their availability, and minimising downtime.

Furthermore, IoT in asset management extends beyond physical assets. It also encompasses the tracking and management of digital assets, such as cryptocurrencies. IoT devices securely monitor and manage digital wallets, providing users with real-time updates on their holdings and transactions. This transparency and control can instil confidence in users and contribute to the overall stability of the financial ecosystem.

While the benefits of implementing IoT in finance are significant, there are several challenges that financial institutions need to address to ensure successful adoption. One of the main challenges is data security and privacy. With the increased connectivity of IoT devices, there is a greater risk of cyberattacks and data breaches. It is crucial for financial institutions to implement robust security measures, such as encryption and authentication protocols, to protect sensitive customer information.

Another challenge is the interoperability of IoT devices. As the number of IoT devices increases, it is essential for these devices to communicate and share data seamlessly. Standardisation of protocols and data formats is necessary to ensure interoperability, enabling different devices to work together and exchange information effectively.

Moreover, the scalability of IoT infrastructure is another challenge that financial institutions need to consider. As the number of IoT devices grows, the infrastructure supporting them must be able to handle the increased volume of data and transactions. Financial institutions need to invest in scalable and reliable IoT platforms to ensure smooth operations and avoid system failures.

The field of IoT in finance is constantly evolving, with new innovations being introduced regularly. One of the recent innovations is the use of blockchain technology in IoT for finance. Blockchain, known for its decentralised and immutable nature, can enhance the security and transparency of IoT transactions. By combining blockchain and IoT, financial institutions can ensure secure and trustworthy transactions, reducing the risk of fraud and tampering.

Another innovation is the development of edge computing in IoT for finance. Edge computing refers to the processing and analysis of data at the edge of the network, closer to the source of data generation. This enables real-time data analysis and decision-making, reducing latency and improving overall system performance. In the finance industry, edge computing can be used for high-frequency trading, fraud detection, and real-time risk assessment.

Moreover, the integration of artificial intelligence (AI) and machine learning (ML) with IoT is another area of innovation. AI and ML algorithms can analyse vast amounts of data collected from IoT devices, identifying patterns, detecting anomalies, and making predictions. This enables financial institutions to automate processes, personalise services, and make data-driven decisions, leading to greater efficiency and profitability.

As financial institutions adopt IoT technologies, it is crucial to consider the *regulatory landscape* to ensure compliance and protect the interests of customers. With the increased collection and processing of personal and financial data, institutions must adhere to data protection regulations, such as the General Data Protection Regulation (GDPR). Financial institutions need to implement robust data privacy and security measures to protect customer information and ensure compliance with regulatory requirements.

Additionally, financial institutions need to address the ethical implications of the IoT in finance. The use of IoT devices raises concerns about privacy invasion and surveillance. It is important for financial institutions to be transparent about the data they collect, how it is used, and obtain proper consent from customers. Institutions should also establish clear policies and guidelines for the ethical use of IoT data to build trust with customers and maintain a positive reputation.

To successfully implement IoT in a finance organisation, it is essential to follow a structured approach. Here are the key steps to consider:

Identify Business Objectives: Define the goals and objectives that IoT implementation aims to achieve. This could include improving operational efficiency, enhancing customer experience, or reducing costs.

Conduct a Feasibility Study: Assess the technical and financial feasibility of implementing IoT in the organisation. Consider factors such as infrastructure requirements, budget, and potential risks.

Develop a Roadmap: Create a detailed plan that outlines the steps and timeline for IoT implementation. This should include identifying the necessary IoT devices, data collection methods, and integration with existing systems.

Select IoT Vendors: Research and select reliable IoT vendors that provide suitable devices and platforms for the organisation's needs. Evaluate factors such as security, scalability, and compatibility with existing technologies.

Integrate IoT Devices: Install and configure IoT devices according to the organisation's requirements. Ensure proper connectivity and establish protocols for data collection, transmission, and storage.

Implement Data Analytics: Develop data analytics capabilities to process and analyse the data collected from IoT devices. This can involve the use of AI and ML algorithms to gain valuable insights and make data-driven decisions.

Monitor and Evaluate: Continuously monitor the performance and effectiveness of IoT implementation. Collect feedback from employees and customers to identify areas for improvement and address any issues that arise.

The IoT has the potential to revolutionise the finance industry, driving transformation and innovation. By leveraging IoT technologies, financial institutions can enhance operational efficiency, improve customer experience, and provide innovative products and services. However, successful implementation of the IoT in finance requires addressing challenges such as data security, interoperability, and scalability.

As the field of IoT in finance continues to evolve, financial institutions need to stay updated with the latest innovations and regulatory requirements. By following a structured approach and considering the steps outlined, organisations can successfully implement IoT and unlock its full potential. Embracing the future with IoT in finance will enable financial institutions to stay competitive, meet the changing needs of customers, and drive transformation and innovation in the industry.

CHAPTER 18

Data Science, Analytics, and Finance Data Management

I N TODAY'S DIGITAL AGE, data has become the fuel that power businesses across all sectors. The sheer volume, variety, and velocity of data available have opened up new possibilities for companies to gain valuable insights and make informed decisions. As businesses embrace the potential of big data, the role of data science and analytics has become paramount in driving finance transformation and enabling leaders to navigate the complexities of an increasingly data-driven world.

Wang (2024) argues data science offers the greatest immediate potential to accelerate finance transformation. It can bring significant improvements at little cost. However, Wang also acknowledges the security risks posed by data science and the use of technology to gather vast swathes of information and data for calculation and transmission. Wang accepts it is extremely easy to cause security risks such as information leakage and theft (Wang, 2024).

Data serves as the foundation for machine learning (ML) algorithms, enabling businesses to train models and make accurate predictions. ML algorithms, such as photo recognition, require massive datasets to identify patterns and learn from them. The exponential growth of available data is fuelled by an expanding list of sources, including traditional databases, cloud-based software systems, the internet, and social media platforms. Businesses that harness this wealth of data gain a competitive advantage, while those without a plan risk drowning in the data deluge.

Data science plays a crucial role in automating processes and driving digital transformations within organisations. By pulling information from raw data, data scientists extract meaningful insights that can shape business strategies and operations. Data science incorporates components of statistics, math, data engineering, ML, deep learning, and visualisation. Data scientists with expertise in these areas can uncover valuable patterns, detect fraudulent behaviour, recommend personalised content, and optimise business processes.

Data science is the science of transforming raw data into meaningful information. Just as oil is extracted from the ground, data science extracts insights from vast amounts of data.

DOI: 10.1201/9781003514503-18

This process involves capturing data from various sources, maintaining data integrity through effective information governance, processing data through techniques like data mining and ML algorithms, and communicating the results through data reporting and visualisation. *Data analysis* adds value by finding greater meaning in the data, enabling businesses to make data-driven decisions.

Data science involves several key components that contribute to its effectiveness:

Capturing Data: Data is collected from internal and external sources, including databases, websites, and IoT devices, to fuel analytics and decision-making.

Maintaining Data: Effective information governance and data architecture ensure the integrity and accessibility of data throughout its lifecycle.

Processing Data: Data mining, ML algorithms, and other techniques are used to transform raw data into actionable insights.

Communicating Results: Data reporting and visualisation techniques help stakeholders understand and interpret the findings.

Analysing Results: By analysing results, organisations can gain deeper insights into user behaviour, market trends, and future outcomes.

Data science teams comprise three main roles: data analysts, data engineers, and data scientists. *Data analysts* are responsible for gathering and organising data, spending a significant amount of time cleaning and preparing the data for analysis. *Data engineers* build the tools and infrastructure to support data scientists' work, integrating data from various sources and ensuring its accessibility. *Data scientists* leverage their expertise in statistics, math, ML, and domain-specific knowledge to analyse data, uncover insights, and make predictions.

Business leaders, particularly Chief Finance Officer (CFO), play a critical role in building and managing data science teams. CFOs need to understand the differences between data analysts, data engineers, and data scientists to effectively integrate data into their everyday business operations. By embracing data-driven decision-making, CFOs can combine their financial expertise with data insights to drive strategic initiatives, improve financial planning and analysis, and gain a competitive edge.

Let's consider the case study of a regional Italian restaurant chain in the United Kingdom. AB, the owner, leverages data science to determine the best locations for opening new restaurants. By analysing data on competition, customer demographics, population density, income levels, and other factors, AB can make informed decisions about site selection and predict the success of a new restaurant. This case study exemplifies how data science can empower business owners to drive growth and make data-backed decisions.

As businesses embrace digitisation and data analytics, the role of the CFO is transforming. CFOs are no longer just 'bean counters' but rather strategic leaders who combine financial expertise with data-driven insights. By understanding the power of data science, CFOs can make informed predictions, identify trends, mitigate risks, and drive business growth. This shift in mindset from relying on hunches to embracing data-driven hypotheses enables CFOs to guide their organisations towards success in the age of analytics.

In my experience, to fully leverage data science and analytics, CFOs must embrace technology and automation. By adopting advanced finance software, CFOs can automate transactional tasks, reduce errors, and shift their focus to strategic initiatives and advanced analytics. Additionally, CFOs can establish a data-driven culture by working with IT to ensure data availability across all departments, implementing data analytics and visualisation techniques, and delivering actionable information to key decision-makers. Embracing technology expands the CFO's role, allowing them to lead their organisations to success in the digital era.

Finance transformation in the age of analytics relies heavily on data science and analytics. By harnessing the power of data, businesses can make informed decisions, optimise operations, and drive growth. CFOs play a crucial role in this transformation, evolving from traditional financial specialists to data-driven leaders who combine financial expertise with data insights. By embracing technology and building data science teams, CFOs can lead their organisations towards success in the dynamic and data-rich business landscape.

In today's fast-paced business environment, finance transformation has become a critical aspect for companies seeking to stay competitive. At the heart of this transformation lies the power of data scientists. These highly skilled professionals play a pivotal role in leveraging the vast amount of financial data to drive strategic decision-making and improve business outcomes.

Data scientists in finance transformation are responsible for analysing complex financial data sets to uncover valuable insights that can inform key business strategies. They possess a unique blend of technical expertise and financial acumen, enabling them to navigate through large datasets and extract meaningful patterns. By identifying trends, predicting market behaviours, and detecting anomalies, data scientists empower organisations to make data-driven decisions that can lead to increased profitability and sustainable growth.

Furthermore, data scientists bring a fresh perspective to finance transformation initiatives. Their ability to think critically and creatively allows them to challenge traditional methods and uncover innovative solutions. By leveraging advanced analytics and ML techniques, data scientists can identify new revenue streams, optimise financial processes, and mitigate risks. Their contribution to finance transformation extends beyond number crunching, as they bring a holistic approach that combines statistical analysis with business acumen.

Finance data management is the backbone of any successful finance transformation initiative. It involves the collection, storage, and analysis of financial data to derive actionable insights. Effective finance data management enables organisations to have a single source of truth, ensuring accuracy and consistency in financial reporting. This allows decision-makers to have timely and reliable information at their fingertips, enabling them to make informed decisions that drive business growth.

In the context of finance transformation, data management becomes even more crucial. As organisations strive to leverage data as a strategic asset, it is essential to have a robust data management framework in place. This involves implementing data governance practices, establishing data quality standards, and ensuring data privacy and security. By having a strong foundation in finance data management, organisations can unlock the true potential of their data scientists and maximise the impact of finance transformation initiatives.

Data scientists play a pivotal role in driving the success of finance transformation initiatives. Their unique skillset and expertise enable them to extract valuable insights from vast amounts of financial data, empowering organisations to make data-driven decisions. Here are some ways in which data scientists can drive success in finance transformation:

Predictive Analytics: Data scientists can leverage predictive analytics techniques to forecast market trends, customer behaviour, and financial outcomes. By identifying patterns and correlations within financial data, they can provide valuable insights that inform strategic decision-making.

Risk Management: Data scientists can help organisations identify and mitigate financial risks by analysing historical data and developing risk models. By identifying potential risks and providing recommendations for risk mitigation, data scientists can enhance the overall risk management framework.

Process Optimisation: Data scientists can optimise financial processes by analysing data and identifying areas of improvement. By automating manual tasks, streamlining workflows, and identifying bottlenecks, data scientists can drive efficiency and reduce costs.

Revenue Optimisation: Data scientists can help organisations identify new revenue streams and optimise existing ones. By analysing customer data, market trends, and financial metrics, they can uncover opportunities for revenue growth and develop strategies to capitalise on them.

To excel in the field of finance transformation, data scientists need to possess a specific set of skills and qualifications. Here are some key skills and qualifications that are essential for data scientists in finance:

Strong Quantitative Skills: Data scientists need to have a strong foundation in mathematics, statistics, and econometrics. They should be proficient in statistical analysis, predictive modelling, and optimisation techniques.

Domain Knowledge: Data scientists in finance should have a deep understanding of financial markets, products, and regulations. They should be familiar with financial statements, valuation techniques, and risk management concepts.

Programming Skills: Proficiency in programming languages such as Python, R, and Structured Query Language (SQL) is essential for data scientists in finance. They should be able to write efficient code, manipulate large datasets, and implement ML algorithms.

Data Visualisation: Data scientists should be able to effectively communicate their findings through data visualisation techniques. They should be proficient in tools such as Tableau, Power BI, or Matplotlib to create visually appealing and informative dashboards.

Business Acumen: Data scientists need to have a strong business acumen to understand the strategic goals and challenges of the organisation. They should be able to translate data insights into actionable recommendations that align with the overall business objectives.

While data scientists play a crucial role in finance transformation, they also face several challenges and opportunities. These challenges stem from the unique nature of financial data and the evolving landscape of finance transformation. Here are some challenges and opportunities for data scientists in finance transformation:

Data Quality and Availability: Financial data is often complex, fragmented, and prone to errors. Data scientists need to ensure data quality and integrity to derive accurate insights. They also need to address data availability issues, as financial data is often scattered across multiple systems and sources.

Regulatory Compliance: Data scientists in finance transformation need to navigate through complex regulatory frameworks and ensure compliance with data privacy and security regulations. They need to develop robust data governance practices and implement strict security measures to protect sensitive financial information.

Technological Advancements: The field of data science is constantly evolving with new tools, algorithms, and technologies. Data scientists need to stay updated with the latest advancements in ML, artificial intelligence, and big data analytics to leverage these technologies in finance transformation initiatives.

Collaboration and Communication: Data scientists need to collaborate with various stakeholders, including finance professionals, IT teams, and senior management. Effective communication skills are essential to bridge the gap between technical expertise and business requirements.

Despite these challenges, data scientists in finance transformation have ample opportunities to make a significant impact. By leveraging advanced analytics and ML techniques, they can unlock the true potential of financial data and drive strategic decision-making. They have the opportunity to shape the future of finance by transforming traditional finance processes and driving innovation.

Data scientists in finance employ a wide range of *tools and technologies* to analyse and derive insights from financial data. These tools enable them to process large datasets, develop predictive models, and visualise data in a meaningful way. Here are some commonly used tools and technologies by data scientists in finance:

Python: Python is a popular programming language used by data scientists for its versatility and extensive library ecosystem. It offers powerful libraries such as NumPy, Pandas, and scikit-learn, which facilitate data manipulation, statistical analysis, and ML.

R: R is another programming language widely used in data science, particularly in statistical analysis and visualisation. It offers a rich set of packages for data manipulation, exploratory data analysis, and modelling, making it a preferred choice for data scientists in finance.

ML Libraries: Data scientists in finance leverage ML libraries such as TensorFlow, Keras, and PyTorch to develop predictive models and automate financial processes.

SQL: SQL is essential for data scientists in finance to extract and manipulate data from relational databases. It enables them to write queries to retrieve specific data points and perform aggregations for analysis.

Tableau: Tableau is a powerful data visualisation tool that allows data scientists to create interactive dashboards and reports. It enables them to present complex financial data in a visually appealing and easily understandable format.

Apache Hadoop: Apache Hadoop is a framework used for the distributed processing of large datasets. It enables data scientists to perform parallel processing and handle big data analytics efficiently.

To ensure the successful integration of data science in finance transformation, organisations should follow best practices that optimise the impact of data scientists. Here are some best practices for integrating data science in finance transformation:

Define Clear Objectives: Clearly define the objectives and expected outcomes of the finance transformation initiative. This will help data scientists align their efforts with the overall business goals and focus on delivering tangible results.

Collaboration between Finance and IT: Foster collaboration between finance professionals and IT teams. This collaboration ensures that data scientists have access to the necessary data, tools, and infrastructure to perform their analysis effectively.

Invest in Data Governance: Implement a robust data governance framework to ensure data quality, privacy, and security. This includes establishing data standards, defining data ownership, and implementing data cleansing and validation processes.

Continuous Learning and Development: Encourage data scientists to stay updated with the latest advancements in data science and finance. This can be done through training programmes, conferences, and knowledge-sharing sessions.

Promote a Data-Driven Culture: Foster a data-driven culture within the organisation by promoting the use of data in decision-making at all levels. This involves creating awareness about the value of data and providing access to data and analytics tools to relevant stakeholders.

To illustrate the power of data scientists in finance transformation, let's look at some real-world case studies:

Fannie Mae: Fannie Mae, a financial services firm, leveraged data scientists to identify patterns in customer behaviour and develop personalised investment portfolios. By analysing vast amounts of customer data, they were able to offer tailored investment recommendations, resulting in increased customer satisfaction and loyalty.

Kroger: Kroger, a multinational retailer, used data scientists to optimise their supply chain and inventory management processes. By analysing sales data, market trends, and historical demand patterns, they were able to reduce stockouts, improve inventory turnover, and optimise their supply chain network.

Generali Group: Generali Group, a leading insurance provider, employed data scientists to develop risk models and predict fraudulent claims. By analysing historical claims data and identifying patterns of fraudulent behaviour, they were able to reduce losses and improve the accuracy of their underwriting processes.

These case studies demonstrate the transformative power of data scientists in finance. By leveraging data analytics and advanced modelling techniques, organisations can make better-informed decisions, optimise financial processes, and drive business growth.

Finance transformation is no longer just a buzzword; it has become a strategic imperative for organisations across industries. In this digital age, the power of data scientists is crucial in unlocking the true potential of financial data and driving meaningful insights. By leveraging their unique skill set and expertise, data scientists can help organisations make data-driven decisions that lead to improved financial performance and sustainable growth.

However, for data scientists to drive success in finance transformation, organisations need to provide them with the necessary tools, resources, and support. This includes investing in robust data management practices, fostering collaboration between finance and IT teams, and promoting a data-driven culture. By doing so, organisations can unleash the full potential of data scientists and embark on a transformative journey that revolutionises their finance processes.

Embrace the power of data scientists in finance transformation and unlock a world of opportunities for your organisation. Start harnessing the power of financial data and make data-driven decisions that propel your business forward.

CHAPTER 19

Cybersecurity During Transformation

I n today's digital world, where technology is rapidly advancing, the importance of cybersecurity cannot be overstated. With businesses relying heavily on digital systems and processes, the risk of cyber threats has become a major concern. This is especially true in the finance industry, where sensitive financial information and transactions are at stake. CFOs must consider the significance of cybersecurity in driving finance transformation and how it plays a vital role in securing business operations.

Finance transformation is an important process of reimagining and redesigning financial systems, processes, and operations to improve efficiency, reduce costs, and enhance decision-making capabilities. It involves implementing new technologies, such as automation, artificial intelligence, and data analytics, to streamline financial processes and enable real-time insights.

The impact of finance transformation on businesses is profound. It enables organisations to make data-driven decisions, improve financial forecasting accuracy, and optimise resource allocation. It also promotes collaboration across departments and enhances the overall agility and responsiveness of the finance function. However, with these advancements come new vulnerabilities that need to be addressed through robust cybersecurity measures.

Avira et al. (2023) contend that cybersecurity and the privacy of financial data are the most important challenges to be addressed in digital transformation in financial management (Avira et al., 2023). This becomes a pivotal risk to be addressed, especially when incorporating emerging technologies.

In my experience, cybersecurity plays a crucial role in driving finance transformation by safeguarding critical financial data, protecting against cyber threats, and ensuring regulatory compliance. It provides the necessary security framework to enable organisations to embrace digital transformation without compromising their financial integrity.

DOI: 10.1201/9781003514503-19

Wang (2023) argues cybersecurity remains a key risk in digital transformation in finance and organisations need to improve risk management to implement appropriate precautions to protect customer data (Wang, 2023).

One of the primary roles of cybersecurity in finance transformation is to protect sensitive financial information from unauthorised access, theft, or manipulation. With the increasing sophistication of cyberattacks, organisations must implement strong security measures, such as encryption, multi-factor authentication, and intrusion detection systems, to prevent data breaches and financial fraud.

Cybersecurity helps organisations comply with industry regulations and standards. The finance industry is heavily regulated, with strict requirements for data protection and privacy. In my experience, by implementing robust cybersecurity measures, organisations can ensure compliance with these regulations and avoid costly penalties or reputational damage.

While cybersecurity is crucial in finance transformation, it also presents several challenges that organisations must address. One of the main challenges is the ever-evolving nature of cyber threats. Cybercriminals are constantly developing new techniques and strategies to exploit vulnerabilities in financial systems. This requires organisations to stay updated with the latest cybersecurity technologies and practices to effectively mitigate these threats.

Another challenge is the shortage of skilled cybersecurity professionals. The demand for cybersecurity expertise is growing rapidly, but there is a shortage of qualified professionals to meet this demand. Organisations must invest in training and development programmes to build a strong cybersecurity workforce and ensure the effective implementation of cybersecurity measures.

Additionally, the complexity of financial systems and processes can pose challenges in implementing cybersecurity measures. Finance transformation often involves integrating multiple systems and technologies, which can create vulnerabilities if not properly secured. Organisations must conduct thorough risk assessments and develop comprehensive cybersecurity strategies to address these challenges.

To overcome the cybersecurity challenges in finance transformation, organisations should follow best practices to ensure the effective implementation of cybersecurity measures.

Further advice on cybersecurity best practices is provided in the next section but here are some key practices to consider:

Develop a Comprehensive Cybersecurity Strategy: Organisations should develop a well-defined cybersecurity strategy that aligns with their finance transformation goals. This strategy should include risk assessments, incident response plans, and regular security audits to identify and address vulnerabilities.

Implement Strong Access Controls: Access controls play a crucial role in protecting financial systems and data. Organisations should implement strong authentication mechanisms, such as multi-factor authentication, to ensure that only authorised personnel can access sensitive financial information.

Regularly Update and Patch Systems: Keeping systems and software up to date is essential in mitigating cybersecurity risks. Organisations should regularly update and patch their financial systems to address any known vulnerabilities and protect against emerging threats.

Provide Cybersecurity Training and Awareness: Human error is one of the leading causes of cybersecurity breaches. Organisations should provide regular cybersecurity training and awareness programmes to educate employees about potential threats, safe online practices, and the importance of reporting suspicious activities.

Partner with Cybersecurity Experts: Engaging cybersecurity experts can provide organisations with specialised knowledge and expertise in implementing effective cybersecurity measures. These experts can help organisations navigate through complex security challenges and ensure the success of finance transformation initiatives.

Several organisations have successfully implemented finance transformation initiatives with a strong cybersecurity foundation. One such example is *Lloyds Bank*, which leveraged advanced data analytics and automation technologies to streamline its financial processes.

Lloyds Bank also implemented robust cybersecurity measures, including real-time threat monitoring and advanced encryption, to protect its customers' financial data.

Another example is *AXA Insurance*, which underwent a finance transformation journey to enhance its operational efficiency and customer experience. AXA Insurance integrated its financial systems with advanced cybersecurity solutions, such as intrusion detection systems and behaviour analytics, to detect and prevent cyber threats. This enabled AXA Insurance to safeguard customer data while improving its overall financial performance.

These examples demonstrate the importance of incorporating cybersecurity into finance transformation initiatives to ensure their success and protect against potential risks.

As technology continues to evolve, the future of cybersecurity in driving finance transformation looks promising. Organisations will need to adapt to emerging technologies and trends to stay ahead of cyber threats and maintain a secure financial environment.

One emerging trend is the use of artificial intelligence and machine learning in cybersecurity. These technologies can analyse vast amounts of data in real-time to identify patterns and anomalies that indicate potential cyber threats. By leveraging AI and ML, organisations can enhance their cybersecurity capabilities and proactively address emerging risks.

Another trend is the increased focus on cloud security. With the growing adoption of cloud computing in the finance industry, organisations need to ensure the security of their cloud-based financial systems and data. Cloud security solutions, such as data encryption, access controls, and secure APIs, will play a vital role in protecting financial information in the cloud.

In addition to AI, ML, and cloud security, there are several other cybersecurity trends that will shape finance transformation in the future. These include:

Zero Trust Security: Zero trust security is a model that assumes no user or device can be trusted by default. It requires continuous verification of users' identities and devices' security postures before granting access to financial systems. This approach provides an additional layer of security to protect against insider threats and unauthorised access.

Blockchain Technology: Blockchain technology has the potential to revolutionise cybersecurity in the finance industry. Its decentralised and immutable nature makes it highly secure for financial transactions and data storage. Blockchain can enhance the integrity and transparency of financial systems, reducing the risk of fraud and manipulation.

Cyber Threat Intelligence: Cyber threat intelligence involves collecting and analysing data about potential cyber threats to identify trends, patterns, and indicators of compromise. By leveraging threat intelligence, organisations can proactively detect and mitigate cyber threats before they cause significant damage.

These trends highlight the continuous evolution of cybersecurity in finance transformation and the need for organisations to stay updated with the latest technologies and practices.

Partnering with cybersecurity experts can bring numerous benefits to organisations undergoing finance transformation. These experts have specialised knowledge and expertise in cybersecurity, enabling organisations to navigate through complex security challenges and implement effective cybersecurity measures.

By partnering with cybersecurity experts, organisations can:

Leverage Industry Best Practices: Cybersecurity experts stay updated with the latest industry best practices and regulations. They can provide guidance on implementing the most effective cybersecurity measures and ensuring compliance with regulatory requirements.

Mitigate Cybersecurity Risks: Cybersecurity experts can conduct thorough risk assessments and identify vulnerabilities in financial systems. They can develop comprehensive cybersecurity strategies and implement the necessary controls to mitigate cybersecurity risks effectively.

Enhance Incident Response Capabilities: In the event of a cybersecurity breach, cybersecurity experts can help organisations respond quickly and effectively. They can assist in investigating the breach, containing the incident, and restoring normal operations while minimising the impact on business operations.

Cybersecurity plays a vital role in driving finance transformation by safeguarding critical financial data, protecting against cyber threats, and ensuring regulatory compliance. It is essential for organisations to understand the significance of cybersecurity in finance transformation and proactively implement robust cybersecurity measures.

Despite the challenges posed by evolving cyber threats and the shortage of skilled cybersecurity professionals, organisations can overcome these challenges by following best practices and partnering with cybersecurity experts. By embracing cybersecurity as a catalyst for successful finance transformation, organisations can enhance their financial operations, improve decision-making capabilities, and gain a competitive edge in the digital age.

CHAPTER 20

Cybersecurity and Data Privacy Regulations Requirements

CYBERSECURITY INCIDENTS HAVE BECOME increasingly common in the business world, with numerous data breaches exposing billions of records. Among the industries most affected by these threatening cyberattacks is the financial sector. In fact, the vast majority of fintech businesses are vulnerable to severe cyberattacks, including app security attacks, ransomware, and phishing.

Gao (2024) argues cybersecurity remains the biggest hurdle in digital transformation in the financial industry and that there is significant potential for harm to data privacy caused by vulnerabilities impacting digital safety and stability (Gao, 2024).

To hold financial service providers accountable for their security posture, cybersecurity regulations have been implemented. Compliance with these regulations ensures that businesses adhere to key security requirements, reducing their vulnerability to security incidents. However, navigating the complex landscape of cybersecurity regulations can be challenging, with numerous intricate details to consider. Chief Finance Officer (CFO) and Chief Information Officer (CIO) must become more conversant with cybersecurity regulations in the financial services industry to ensure compliance and implement associated best practices to reduce their cyber risk.

Financial cybersecurity compliance refers to the security regulations implemented by financial institutions to prevent data breaches and maintain a strong security posture. These regulations align with the laws and security requirements that provide the minimum standard for data protection within the financial industry. Financial cybersecurity compliance impacts various sectors within the financial services industry, including mutual funds, investment banks, commercial banks, brokerage firms, insurance companies, credit unions, and wealth management firms.

The financial services industry faces unique challenges when it comes to cybersecurity compliance due to the variety of security standards and the overlaps between them. However, focusing on the regulations specific to financial institutions can help overcome this challenge. Here are the main cybersecurity regulations that financial services companies must adhere to, each promoting customer data security and data breach resistance.

20.1 PAYMENT CARD INDUSTRY DATA SECURITY STANDARDS (PCI DSS)

The Payment Card Industry Data Security Standards (PCI DSS) is a set of security standards developed to reduce credit card fraud and secure the sensitive information of credit cardholders. Compliance with PCI DSS is mandatory for organisations that receive or process customer credit card information. The standards include technical and operational requirements developed and managed by the PCI Security Standards Council.

20.2 REQUIREMENTS OF PCI DSS

The most important criterion of PCI compliance is that a company must secure other people's payment information as carefully as they would their own. The firm must document how it is mitigating cybersecurity or data privacy risks such as accidental leakage of credit card information or losing documents containing customer personal information. Every firm must safeguard its client transaction history, account information, and personal information. PCI DSS compliance requirements assist firms in adhering to secure business practices and ensuring that their customers are adequately protected.

20.3 SARBANES-OXLEY ACT

The Sarbanes-Oxley (SOX) Act was enacted in the USA to protect investors against financial fraud. The SOX framework provides recommended security procedures for avoiding fraudulent financial activities. Compliance with SOX is essential for all publicly listed businesses, including those in the financial sector. The act mandates corporate governance and financial transparency, requiring accurate representation of a company's financial facts and comprehensive reviews of policies, controls, and procedures. SOX has grown beyond the framework and provides cybersecurity sections to provide greater assurance that financial institutions are prepared to deal with cybersecurity threats that could disrupt financial transactions or undermine the security of financial records.

20.4 REQUIREMENTS OF SOX

The SOX Act's provisions cover corporate governance and financial transparency and are a requirement for companies based in the United States and for non-US-based businesses. This act means there are obligations that all financial reports include an internal controls report. Internal controls reports must be an accurate representation of a company's financial facts. In an audit of Section 404, an auditor at SOX must systematically review policies, controls, and procedures, providing assurance that internal controls and processes can be audited using a control framework.

20.5 ISO/IEC 27001

ISO/IEC 27001 is an internationally recognised standard for lowering security risks and safeguarding information systems. While ISO 27001 certification is not mandatory in most countries, it is highly recommended for businesses in the financial services sector due to its superior protection of sensitive data. The requirements include scoping an Information Security Management System (ISMS), conducting a risk assessment, and implementing a risk treatment methodology.

Financial service firms that will not follow ISO 27001 can still improve cybersecurity by following the list of domains and controls provided. Abiding by this framework also benefits your organisation with General Data Protection Regulation (GDPR) compliance when implemented along with an ISMS.

20.6 REQUIREMENTS OF ISO 27001

During the implementation of ISO/IEC 27001, the most important steps are scoping your ISMS (defining what information needs to be protected), completing a risk assessment, and creating a risk treatment methodology (identifying threats to your information).

Obligatory clauses must also be completed by organisations, such as the risk treatment plan, risk assessment report, and information security policy and objectives. Also, obligatory are the information risk treatment process, the internal audit programme, the results of internal audits, and the records of skills, qualifications, training, and experience. Other clauses which are obligatory are the results of the management review, the results of corrective actions, and the monitoring and measurement of results.

20.7 NATIONAL INSTITUTE OF STANDARDS AND TECHNOLOGY (NIST)

The National Institute of Standards and Technology (NIST) provides a wide range of information security requirements, including cybersecurity compliance. It can be considered the American equivalent of the ISO standards. NIST has a wide range of information security requirements, including cybersecurity compliance, which is addressed in NIST document 800-53. NIST 800-53 revision 5 widened the scope to cover non-government bodies and contains a single set of controls to facilitate the harmonisation of numerous standards, in addition to a stronger emphasis on data security than prior revisions. While compliance with NIST is mandatory for US federal entities and their contractors, it is voluntary for private sector businesses, including financial service providers. NIST outlines 110 requirements covering various aspects of an organisation's IT technology, procedures, and policies.

20.8 REQUIREMENTS OF NIST

NIST has a group of 110 requirements for an organisation's IT technology, procedures, and policies. Access control, system configuration, and authentication methods are all covered by these requirements. Cybersecurity protocols and incident response plans are also important areas covered. Each requirement deals with a cybersecurity vulnerability or improves a network component, and it comes with a detailed explanation that helps the organisation understand the wider context. Implementation of each requirement ensures an organisation's network, systems, and employees are adequately prepared to securely handle any controlled Unclassified Information.

20.9 GENERAL DATA PROTECTION REGULATION

The GDPR is a European Union security framework designed to prevent the compromise of personal data. Compliance with GDPR is required for financial services that collect or process personal data from EU residents. The regulation emphasises the protection of personal data throughout its lifecycle and includes principles such as limited purpose, fairness, transparency, limitation on storage, accuracy, integrity and confidentiality, and data minimisation.

20.10 REQUIREMENTS OF GDPR

Organisations are mandated to ensure data protection measures are implemented to protect consumers' personal data and privacy from loss or exposure. The most significant principles and requirements govern the management of personal data and are summarised in Article 5 of the GDPR. These include:

- Fairness, Lawfulness, and Transparency: Personal data should be processed in a way that is legitimate, fair, and transparent.

- Limited Purpose: The personal data of customers should only be gathered for legitimate, explicit, and specific objectives and should never be processed in a way that is incompatible with these goals.

- Limitation on Storage: Personal data should not be stored for any longer than is required for the purposes for which it is processed.

- Data Minimisation: Personal data gathering should be minimised, and data acquired must be useful to achieve a defined goal.

- Accuracy: Personal data should be accurate and, where appropriate, kept up to date when stored and managed.

- Integrity and Confidentiality: Personal data should be treated in a way that ensures proper security, such as protection against unauthorised or unlawful processing, as well as accidental loss, destruction, or damage.

20.11 DIGITAL OPERATIONAL RESILIENCE ACT (DORA)

DORA is a new general legislative framework adopted by the European Commission for digital operational resilience in the financial sector. It was formally adopted in January 2023 and will come into force in January 2025. DORA applies to financial entities, such as credit institutions, investment firms, payment service providers, insurance and reinsurance undertakings, asset managers, crypto-asset service providers, and market infrastructures ('Financial Entities (FE)(s)'),[2] and companies that provide Information and communication technology (ICT) services to such entities (ICT service providers), establishing a common set of rules and standards for:

- Management of ICT risks

- Oversight of critical ICT third-party providers

- Reporting and information sharing of ICT incidents

- Testing and auditing of ICT systems and processes in the context of an FE's business

The rules and standards established by DORA are to be implemented through several EU Commission-delegated regulations, to 'bring to life' the general principles and high-level requirements set forth by DORA. The Act has a similar approach to GDPR. In both cases, rather than defining precise rules of conduct, they set forth the general principles that should drive FEs (or data controllers/processors) in ensuring the resilience of their ICT systems and the protection of personal data.

DORA complements the GDPR and shares the same goals: ensuring the security, confidentiality, and integrity of (personal and non-personal) data and the rights and freedoms of data subjects in the digital environment.

DORA will soon have a significant impact on FEs and ICT service providers as it will strengthen ICT requirements and security standards. Organisations should map the obligations DORA will impose on them and do a gap analysis and remediation plan, adopting a holistic approach that considers other laws and regulations providing for security requirements and/or data protection.

20.12 REQUIREMENTS OF DORA

DORA covers core requirements to ensure ICT resilience by financial bodies. These provide a comprehensive digital resilience framework. The key features are:

ICT risk management

a. Setting up resilient ICT systems and tools that minimise the impact of ICT risk.

b. Adopting extensive business continuity policies and disaster and recovery plans.

c. Implementing mechanisms to gather information from both external events and the FE's own ICT incidents to improve the ICT risk management framework.

ICT-related incident reporting

a. Implementing a management process to monitor and log ICT-related incidents.

b. Classifying ICT incidents in accordance with the criteria that will be specified by the applicable supervisory authority, e.g., depending on the business industry, The European Banking Authority (EBA), European Insurance and Occupational Pensions Authority (EIOPA), etc.

c. Reporting ICT incidents to the relevant authorities (and/or to the FEs' users) using a common template and harmonised procedure.

Digital operational resilience testing

a. Periodic testing of the ICT risk-management framework to detect any weakness or deficiency. Such criticalities must be promptly resolved or mitigated by the implementation of additional security measures.

b. Adopting digital operational resilience requirements proportionate to the entities' size, business, and risk profiles.

ICT third-party risk

a. Sound monitoring of risks deriving from the ICT services outsourced to ICT third-party service providers.

b. Ensuring that the agreements with the ICT third-party providers contain all necessary monitoring and accessibility arrangements, such as a full-service level description, indication of locations where data are being processed, etc.

c. Developing a consistent regulatory approach by the different supervisory authorities.

Information sharing

a. Sharing relevant information within the context of other trusted FEs to

i. enhance FEs' digital operational resilience,

ii. raise individual and collective awareness on ICT risks and threats,

iii. minimise ICT threats' ability to spread.

b. Enhancing FEs' defensive and detection techniques, mitigation strategies, or response and recovery stages, including developing an effective training programme for FE staff and management.

c. Entering into information-sharing arrangements to disclose to FEs' trusted communities (e.g., sectorial cyber-response teams, etc.) cyber-threat information and intelligence.

To ensure compliance, it is important to conduct an ICT and privacy risk assessment. Thereafter, appropriate ICT policies and procedures must be maintained to ensure the security and resilience of FE's ICT systems and functions. It is then important to implement an ICT privacy oversight mechanism. Lastly, the financial body must ensure ICT and staff maintain privacy risk awareness.

20.13 GRAMM-LEACH-BLILEY ACT

The Gramm-Leach-Bliley Act (GLBA) requires financial institutions to secure consumer data and disclose data-sharing practices to clients. Compliance with the GLBA is mandatory for all businesses selling financial products or services in the United States. The act includes the Financial Privacy Rule, which mandates privacy notices, and the Safeguards Rule, which requires written information security plans and risk analysis.

20.14 REQUIREMENTS OF GLBA

GLBA has important guidelines covering the protection of personal information or personally identifiable information held by consumers (PII). The key rules include:

- Financial Privacy: This mandates financial organisations to provide privacy notices on the commencement of customer relationships and annually thereafter. This must clarify which information is gathered, where information is shared, how is it utilised, and regarding safeguarding. This must incorporate consumers' rights under the Fair Credit Reporting Act to opt-out of having their personal information shared with any unaffiliated third parties. Unaffiliated parties who receive non-public information (NPI) are bound by acceptance terms under the original relationship agreement.

- Safeguards: This mandates the creation of a written information security plan outlining methods and procedures for safeguarding client NPI. Organisations must undertake a risk analysis of each department that handles NPI and establish, monitor, and test the programme, ensuring data is protected. If there is any change in data management, the protections must be changed as well. The federal government established guidelines for protecting client information.

20.15 PAYMENT SERVICES DIRECTIVE

The Payment Services Directive 2 (PSD 2) is a European Union directive that promotes competition in the banking sector. Compliance with PSD 2 is mandatory for all banks and financial institutions in the EU. The directive includes standards for securing online payments, strengthening customer data security, and implementing strong client authentication.

20.16 REQUIREMENTS OF PSD 2

This regulation has clear requirements to share account information only with third-party service providers (TPPs) with account holders' authorisation. Customers can thereby access a consolidated view of their bank account information through Account Information Service Providers (AISPs). Customers can also start and process online payments through Payment Initiation Service Providers (PISPs). Account Servicing Payment Service Providers are financial firms that handle customer accounts under PSD 2 (ASPSPs). The ASPSPs managing customer accounts must provide a safe means for TPPs to access customer information with the customer's authorisation. TPPs are divided into two kinds in PSD 2: AISPs and PISPs. Customers can access information from several service providers using AISPs. Customers can make online payments straight from their personal bank accounts via PISPs.

Other Regulatory Requirements:

Bank Secrecy Act (BSA)

The BSA is a US regulation also known as the Currency and Foreign Transactions Reporting Act. Its focus is on preventing money laundering and financing terrorism.

Financial Industry Regulatory Authority (FINRA)

FINRA requires US brokers to be registered and licensed. It includes requirements for protecting customer data from exposure and cyber threats. This is not a complete list of financial industry regulations and standards, and the regulatory landscape is changing rapidly. Financial institutions should research applicable regulations and incorporate their requirements into a corporate cybersecurity strategy.

The financial industry is a prime target for cyberattacks, making adherence to cybersecurity regulations and best practices crucial. Here are some key best practices that financial services companies should implement to enhance their cybersecurity posture and meet regulatory requirements.

Access Management and Zero-Trust: Access management is crucial to financial services' data security and regulatory compliance strategies. Many of the major cyber threats to financial services (such as ransomware and data breaches) exploit compromised credentials and privileged access to achieve their objectives. Additionally, managing access to sensitive customer data is the primary focus of most regulations that impact the financial industry.

Financial services organisations can manage access to their data and systems in various ways. Important components of the access management strategy include:

Cloud Application Security Broker (CASB): CASB solutions monitor and manage access to an organisation's cloud-based applications. As financial services pursue digital transformation and shift core applications to the cloud, CASB solutions are vital to preventing unauthorised access to sensitive customer data.

Multi-Factor Authentication (MFA): MFA requires a user to use a combination of factors (such as a password and a physical token) to authenticate to an account. MFA is commonly mandated for access to customer financial data by regulations such as the PCI DSS.

Privileged Access Management (PAM): 74% of data breaches involve third parties with unnecessary privileged access to a company's systems and data. As financial services expand third-party relationships due to open banking and digital transformation initiatives, PAM solutions are vital to monitoring and managing accounts with elevated access to sensitive systems and data.

Zero-Trust Network Access (ZTNA): Telework makes secure remote access solutions necessary as employees remotely access corporate data and systems. ZTNA solutions can enable financial institutions to manage data security risks and regulatory compliance requirements by providing access to data on a case-by-case basis determined by role-based access controls.

Endpoint and User Security: With the rise of remote work, endpoint security is more important than ever for financial institutions. As employees work from home, devices outside the traditional network perimeter have access to sensitive corporate and customer data and enterprise resources. If these devices become infected with malware, an attacker can exploit their remote access to attack the business directly.

Financial services need solutions that enable them to prevent, detect, and respond to potential infections on their employees' devices. Some applicable cybersecurity solutions include:

Extended Detection and Response (XDR): As malware and other cyberattacks grow more sophisticated, traditional standalone endpoint security solutions are growing less and less effective. XDR solutions are designed to take a more holistic approach to threat detection and remediation, collecting data from multiple sources (endpoints, email, network traffic, etc.) and analysing it to identify these more subtle attacks.

Secure Web Gateway (SWG): Many security threats that employees face come over the Internet. Users may accidentally browse to malicious or infected websites or be directed there by phishing emails. SWGs sit between a user and the Internet and proxy all connections, enabling the organisation to block visits to inappropriate or dangerous sites and to monitor for malicious content.

Securing the Distributed Enterprise: As financial institutions embrace remote work and digital transformation initiatives, their IT infrastructure becomes more distributed. This is most obvious in cloud adoption, with the vast majority of financial services organisations shifting to cloud computing.

The shift from on-premises to the cloud creates security challenges for an organisation as traditional perimeter-based defences are no longer effective. Routing all traffic through the headquarters network for security inspection creates network latency and degrades performance, but allowing traffic to continue to its destination uninspected creates risk and risks regulatory non-compliance.

As financial services embrace a more distributed IT infrastructure, they can enhance network performance and security by transitioning to modern security solutions, such as:

Software-Defined WAN (SD-WAN): SD-WAN is a network optimisation tool designed to identify the best route between SD-WAN points of presence over various network media. This can ensure that latency-sensitive applications have the network performance that they require, and security solutions deployed alongside or integrated with SD-WAN solutions can secure all traffic flowing over the enterprise WAN.

Firewall as a Service (FWaaS): Like other cloud-hosted, service-based solutions, FWaaS provides financial services with increased flexibility and scalability. FWaaS can be deployed alongside an organisation's cloud-based applications to protect them and can also be used to provide high-performance and scalable protection to on-premises IT resources.

Threat Detection and Response: Rapid detection and response are vital to minimising the cost and impact associated with a security incident. The longer that an attacker has access to systems, the more opportunity there is to steal or encrypt valuable data, compromise user credentials, or deploy persistence mechanisms to deepen their hold on the system. Financial services need robust threat detection and response to minimise the cost of cybersecurity incidents and meet regulatory deadlines for breach reporting. To do so, companies require visibility into active threats and security personnel capable of addressing them. The following security solutions can help companies meet these requirements.

Security Information and Event Management (SIEM): The average enterprise has 75 security solutions, all of which generate logs and alerts about potential security incidents, overwhelming security analysts with too much data. A SIEM solution aggregates and analyses these logs, using context and multiple data sources to eliminate false positives and draw analysts' attention to the most significant threats.

Managed Detection and Response (MDR): A global cybersecurity skills gap makes it difficult for organisations to find the security talent that they need, leaving 61% of security teams understaffed. MDR can help an organisation fill these gaps by augmenting or replacing in-house security teams with third-party providers with security and compliance expertise.

Regular Vulnerability Assessments and Penetration Testing: Conducting regular vulnerability assessments and penetration testing helps identify and remediate vulnerabilities that could lead to data breaches. These tests also strengthen overall security posture and meet the cyber resilience requirements of regulations.

A zero-trust policy assumes that all network activity is malicious until proven otherwise. This framework promotes secure PAM, making it more difficult for threat actors to gain unauthorised access to critical information.

Improve your Incident Response Plan: Having a well-defined incident response plan enables organisations to effectively respond to security incidents such as data breaches or ransomware attacks. This plan should outline the steps and procedures to be followed in the event of an incident, ensuring a timely and effective response.

Manage Third-Party Risks: Financial institutions often rely on third-party vendors, making it crucial to manage the associated risks. Implementing a robust Third-Party Risk Management (TPRM) solution helps secure the vendor network by certifying cybersecurity improvements and evaluating compliance with security assessments.

Encrypt Valuable Data: Encrypting data is essential for protecting sensitive information and preventing data breaches. By encrypting valuable data, financial institutions can avoid regulatory infractions and associated penalties.

The financial industry faces significant cybersecurity challenges, necessitating compliance with various regulations. Cybersecurity regulations provide a framework for financial services companies to protect customer data and enhance their security posture. By adhering to these regulations and implementing best practices, financial institutions can mitigate cyber risks, safeguard sensitive information, and maintain regulatory compliance. Stay updated with the evolving threat landscape and regulatory changes to ensure the ongoing security of your organisation in the face of cyber threats.

CHAPTER 21

Case Studies of Successful Finance Transformation

21.1 SUCCESSFUL TRANSFORMATION IN FINANCIAL SERVICES

To understand the real-world impact of digital transformation in financial services, let's take a look at some case studies of successful implementations:

Lloyds Bank: Lloyds Bank has embraced digital transformation to enhance customer experiences and drive innovation. They have introduced mobile banking apps, artificial intelligence (AI)-powered chatbots, and biometric authentication. These digital initiatives have improved customer satisfaction and enabled faster and more secure transactions.

PayPal: PayPal disrupted the traditional payment industry with its digital payment platform. They have leveraged digital technologies to offer seamless and secure payment experiences. PayPal's digital transformation has enabled them to expand globally and reach millions of customers worldwide.

Citigroup: Citigroup has invested heavily in digital transformation to stay ahead of the competition. They have implemented AI-powered risk analytics, blockchain for trade finance, and mobile banking apps. These digital initiatives have improved operational efficiency, reduced costs, and enhanced customer experiences.

JPMorgan Chase: JPMorgan Chase, a leading global financial institution, embarked on a comprehensive finance transformation journey to streamline its operations and enhance customer experience. The bank leveraged advanced analytics and AI-powered chatbots to automate customer interactions, provide personalised financial advice, and improve customer satisfaction. By deploying blockchain technology, JPMorgan Chase achieved greater transparency and security in its cross-border transactions while also reducing costs and improving efficiency. The finance transformation initiative enabled the bank to gain a competitive edge, strengthen customer relationships, and drive sustainable growth. In terms of technology investments, the company also stood in first place.[1] According to its annual reports, it invested more than $11.4 billion in AI technology. This massive investment in AI by the topmost leading US bank shows its interest in technological inventions. JPMorgan is majorly focused on how to improve its business performance and achieve operational excellency using AI and robotic process automation (RPA)-like trending banking technologies.

Zurich Insurance: Zurich Insurance, a multinational insurance company, transformed its finance function by adopting cloud-based financial management systems and RPA. By migrating its finance operations to the cloud, Zurich Insurance gained real-time access to critical financial data, improved collaboration among teams, and enhanced reporting capabilities. The company also implemented RPA to automate manual processes such as claim processing and premium calculations. As a result, Zurich Insurance achieved significant cost savings, reduced errors, and improved customer service. The finance transformation initiative enabled the company to streamline its operations, drive efficiency, and deliver value to its customers.

These case studies demonstrate the transformative power of digital technologies in the finance industry. By embracing digital transformation, financial services companies can achieve sustainable growth and create value for their customers.

21.2 REAL-LIFE CASE STUDIES OF FINANCE TRANSFORMATIONS

Finance transformation has become a critical aspect of ensuring a company's long-term success. The ability to adapt and evolve with the changing financial landscape is crucial for businesses to remain competitive. To understand the true power of finance transformation, let's dive into some real-life case studies of successful transformations that have revolutionised companies.

Volkswagen Group: Volkswagen Group is a global manufacturing giant. Facing significant financial challenges due to outdated systems and processes, Volkswagen Group embarked on a finance transformation journey. The company recognised the need to streamline operations, enhance data accuracy, and improve decision-making capabilities. By implementing cutting-edge financial software solutions and reengineering key processes, Volkswagen Group was able to achieve remarkable results. The finance team now has real-time access to accurate financial data, allowing for quicker analysis and decision-making. This transformation has not only improved operational efficiency but has also empowered Volkswagen Group to make informed strategic decisions, resulting in increased profitability and market share.

CitiBank: CitiBank, Facing intense competition in the banking industry, CitiBank realised the need for a comprehensive overhaul of its finance function. The bank's outdated legacy systems and manual processes were impeding growth and hindering customer service.

Through a well-planned finance transformation initiative, CitiBank completely transformed its finance operations. By leveraging advanced technologies like robotic process automation and machine learning, the bank automated mundane tasks, reducing errors and improving efficiency. The implementation of a centralised financial reporting system provided real-time insights, enabling CitiBank to make data-driven decisions. Citibank also made use of AI and RPA-like next-generation technologies, which means they are reaping the benefits of RPA in the banking sector to the fullest. To enhance fraud detection and anti-money laundering, it is investing in AI technology. Recently, Citibank announced its partnership with Feedzai, helping in the detection of fraudulence. This transformation across its finance processes not only enhanced customer experience but also positioned CitiBank as an industry leader, driving significant growth and profitability.

JP Morgan Chase: JP Morgan Chase, a global investment firm, embarked on a finance transformation journey to enhance its operational efficiency and deliver superior investment performance.[2] The company leveraged advanced analytics and machine learning algorithms to automate portfolio management, optimise asset allocation, and identify investment opportunities. J.P. Morgan's Corporate & Investment Bank uses machine learning to personalise the digital experience of its research platform, J.P. Morgan Markets. The platform produces over 10,000 pieces of research a year, but until recently, clients did not always know the reports existed. Machine learning techniques solved the issue, and now each client logs into a customised portal that provides unique and relevant research, personalised to their needs. By implementing cloud-based financial management systems, JP Morgan Chase gained real-time visibility into its financial data, improved collaboration among teams, and enhanced reporting capabilities. The finance transformation initiative enabled the company to achieve significant cost savings, drive innovation, and deliver value to its clients.

Honda: Honda, a leading manufacturer, transformed its finance function by adopting RPA and process automation. By automating manual processes, such as order-to-cash and procure-to-pay, Honda achieved significant cost savings, improved efficiency, and reduced errors.

The company also implemented advanced analytics to gain insights into its financial performance, identify cost-saving opportunities, and optimise working capital. The finance transformation initiative enabled Honda to streamline its operations, enhance decision-making, and drive sustainable growth.

Finance transformations have the potential to revolutionise companies by strengthening financial control, improving decision-making capabilities, and driving sustainable growth. Let's explore a few more real-life case studies that highlight the transformative power of finance.

Nissan: Nissan is a leading company operating in the automotive industry. Faced with challenges like inefficient processes, disparate systems, and a lack of visibility into financial data, Nissan realised the urgent need for change. By implementing an integrated financial management system and streamlining processes, Nissan achieved a significant transformation. The company now has real-time visibility into financial data, allowing for accurate forecasting and budgeting. This newfound visibility has empowered Nissan to optimise cash flow, improve profitability, and drive sustainable growth.

Dropbox: Dropbox is a leading player in the cloud file-sharing industry. Dropbox recognised the need to leverage technology to improve financial control and enhance decision-making capabilities. Using enterprise resource planning (ERP), Dropbox cut its financial period close in half and its accounts receivable period close from four days to one.[3] Cloud Risk Management helped the company automate user access controls to ensure segregation of duties compliance worldwide for all business units. Finance turned to cloud infrastructure to create a simple, secure invoice consolidation process to manage that large volume of monthly transactions. This gave Dropbox direct integration paths with its ERP system and let users access application development tools as a bundle with Oracle Integration, allowing finance to take direct action on the information passing through its systems. Using the prebuilt adapters and low-code automation, Dropbox was able to cut the cost of financial record processing and reduce the volume of transactions while accelerating time to market, making it four times faster. Other finance automation initiatives included a new Intelligent Document Recognition tool to automate the scanning of invoices. Under a big data programme, Dropbox Analytics Cloud, powered by Autonomous Data Warehouse, provides its finance team with access to dashboards, data visualisation tools, and self-service analytics to monitor and improve business performance. Together with Cloud EPM, finance could leverage augmented analytics to manage cash flow, model the impact of new product offerings, and reallocate resources to higher-value initiatives. Dropbox gained real-time insights into its financial performance. This enabled the company to make data-driven decisions, optimise inventory management, and improve profitability. The finance transformation not only resulted in improved operational efficiency but also positioned the company as a leader, driving revenue growth.

21.3 REAL-LIFE CASE STUDIES OF UNSUCCESSFUL FINANCE TRANSFORMATIONS

While successful finance transformations can lead to remarkable outcomes, it is essential to acknowledge that not all transformations yield the desired results. It is crucial to learn from unsuccessful case studies to understand the potential pitfalls and challenges that organisations may face during the transformation process.

One such example is the finance transformation undertaken in 2019 by *Revlon*, a multinational conglomerate.[4] Revlon aimed to streamline its finance function by implementing an ERP system. However, the implementation process was marred by poor planning, a lack of stakeholder engagement, and inadequate training. As a result, the transformation faced significant challenges, leading to disruptions in operations and delayed financial reporting. The lack of proper change management and resistance from employees further exacerbated the situation. Revlon would be late filing its annual financial report because its SAP ERP system had issues. The company's stock fell 6.9% in 24 hours. Revlon's ERP system was implemented in February 2018, not long after they had acquired Elizabeth Arden. This meant the company was balancing a SAP implementation as well as an acquisition. After going live with a new manufacturing plant in North Carolina, problems started to occur. They started to lose customer orders, weren't able to see their supply chain, and were not able to ship products. The failure caused retail sales to drop and operational problems for Revlon. There was also an investor lawsuit made against the company.

Key flaws with the transformation:

- Insufficient risk identification and mitigation.

- There wasn't a clear vision, and all aspects of the project were not sufficiently analysed.

- Effective control measures were not put in place ahead of time.

- They failed to see how organisational change can affect digital transformation success.

Another case study of an unsuccessful finance transformation is the story of *Haribo*, a sweet snack provider. A failure in SAP implementation caused serious supply chain issues for Haribo in 2018, making it impossible to track inventory and raw materials and access inventory at shops. The digital transformation was meant to streamline production and update their goods management system; however, instead, it resulted in a 25% drop in sales.

Key flaws with the transformation:

- Failure to evaluate business processes to ensure they were aligned with the new software.

- Insufficient risk mitigation to plan for things that may go wrong when their SAP went live.

- Failure to ensure SAP implementation met company objectives and goals.

- Insufficient testing phase and checking for controls.

Finance transformation is a critical component of driving sustainable growth in today's business environment. Real-life case studies of successful finance transformations demonstrate the transformative power of advanced technologies, streamlining processes, and leveraging data analytics. However, it is important to learn from unsuccessful case studies to identify potential pitfalls. By measuring the success of finance transformations through key metrics, organisations can continuously improve their finance functions and inspire change throughout the company.

NOTES

1. https://usmsystems.com/robotic-process-automation-in-banking/
2. https://www.jpmorganchase.com/news-stories/tech-investment-could-disrupt-banking
3. https://www.oracle.com/customers/dropbox/
4. https://www.globaldigitalassurance.com/5-high-profile-digital-transformation-fails/

CHAPTER 22

Sustainability

Transformation to a Sustainable Business Model

S USTAINABLE FINANCE HAS BECOME an increasingly important topic in recent years as businesses and individuals recognise the need to address environmental challenges. Chief Finance Officer (CFO) must consider how finance transformation can play a crucial role in driving positive environmental impact and building a sustainable financial future.

Finance is not just about managing money; it can also be a powerful tool for promoting sustainability. By incorporating environmental considerations into financial decision-making processes, organisations can make a significant difference in the world. For example, sustainable finance can involve investing in renewable energy projects, supporting environmentally responsible businesses, or even implementing green lending practices. The choices made by financial institutions and individuals in terms of where they allocate capital can have far-reaching implications for the environment.

Mousa et al. (2023) contend financial businesses that engage in responsible consumption, social sustainability, environmental sustainability, and perceived innovativeness have relative attractiveness, which enhances their offer value and customer satisfaction (Mousa and Bouraoui, 2023).

Integrating sustainability into finance transformation initiatives offers numerous benefits. First and foremost, it aligns financial goals with environmental objectives, creating a win-win situation. By adopting sustainable practices, organisations can enhance their reputation and attract socially conscious investors. Additionally, sustainable finance can drive innovation as it encourages the development of new financial products and services that support environmental goals. Finally, by considering environmental risks and opportunities, organisations can make more informed financial decisions, ultimately leading to long-term value creation.

Numerous organisations have successfully implemented sustainable finance transformation initiatives, demonstrating the positive impact such strategies can have. For instance,

DOI: 10.1201/9781003514503-22

some banks have launched green bond programmes to finance renewable energy projects. These bonds attract investors who are specifically interested in supporting environmentally friendly initiatives. Other companies have embraced sustainable supply chain financing, incentivising suppliers to adopt environmentally responsible practices. By leveraging financial tools and mechanisms, these organisations are driving positive environmental change while also achieving financial success.

To effectively build a sustainable financial future, it is important to follow a set of *key principles*. First, organisations should integrate environmental considerations into their risk management processes, ensuring that potential environmental risks are identified and mitigated. Second, transparency and accountability are crucial. Organisations should disclose their environmental performance and progress towards sustainability goals, enabling stakeholders to make informed decisions. Third, collaboration is essential. Sustainable finance requires partnerships between financial institutions, businesses, and governments to drive change at scale.

Numerous *tools and technologies* are available to support sustainable finance transformation. For instance, there are software solutions that enable organisations to track and analyse their environmental impact, allowing for better decision-making. Additionally, blockchain technology can enhance transparency and traceability in supply chains, ensuring that sustainable practices are upheld. Financial institutions can also leverage artificial intelligence and machine learning to assess environmental risks and identify investment opportunities. These tools and technologies empower organisations to effectively incorporate sustainability into their financial operations.

While there are significant benefits to sustainable finance transformation, there are also challenges and barriers that need to be addressed. One of the main obstacles is the lack of standardised metrics and reporting frameworks for measuring environmental impact. Without consistent and comparable data, it can be challenging for organisations to assess their progress and compare their performance with peers. Additionally, there may be resistance to change within organisations, as sustainable finance transformation often requires a shift in mindset and the adoption of new practices. Overcoming these challenges will require collaboration, education, and the development of industry-wide standards.

Getting started with finance transformation for sustainability can seem overwhelming, but there are practical steps organisations can take. First, it is important to assess the current state of sustainability within the organisation and identify areas for improvement. This could involve conducting a sustainability audit or engaging external consultants. Second, organisations should develop a clear vision and set of goals for their sustainable finance transformation journey. These goals should be specific, measurable, achievable, relevant, and time-bound. Finally, organisations should engage stakeholders, including employees, customers, and investors, to create buy-in and ensure alignment with sustainability objectives.

Finance transformation has the potential to drive positive environmental impact and contribute to the creation of a sustainable future. By incorporating sustainability into financial decision-making processes, organisations can align their financial goals with environmental objectives, attract socially conscious investors, and drive innovation. However, implementing sustainable finance transformation does come with challenges, such as the lack of standardised metrics and resistance to change. Nevertheless, by leveraging tools, technologies, and key principles, organisations can overcome these barriers and make a meaningful difference in the world. It is time for finance to embrace sustainability and play a leading role in building a better future for generations to come.

22.1 THE RISE OF SUSTAINABLE BUSINESS MODELS

In recent years, there has been a significant shift in the way businesses operate. More and more companies are recognising the importance of sustainability and the role it plays in driving profitability and purpose. Sustainable business models are at the forefront of this change, revolutionising the finance industry.

Sustainable business models are strategies that integrate environmental, social, and financial considerations into the core operations of a company. These models aim to create long-term value by aligning business practices with the principles of sustainability. By adopting sustainable business models, organisations can minimise their negative impact on the environment, contribute to social well-being, and enhance their financial performance.

One of the key advantages of sustainable business models is their *positive impact on the environment.* By implementing sustainable practices, companies can reduce their carbon footprint, conserve natural resources, and minimise waste. For instance, many finance organisations are adopting paperless processes, which not only saves trees but also reduces the energy consumption associated with printing and transportation. Moreover, sustainable business models promote the use of renewable energy sources, such as solar and wind power, thus reducing reliance on fossil fuels and mitigating climate change.

Sustainable business models also encourage responsible waste management. Companies are increasingly finding innovative ways to recycle and repurpose materials, reducing the amount of waste sent to landfills. Additionally, sustainable business practices often involve the use of eco-friendly materials and products, which further contribute to environmental preservation. By adopting these practices, finance organisations can not only reduce their environmental impact but also enhance their reputation as socially responsible entities.

In addition to the environmental benefits, sustainable business models also have significant *social advantages.* These models prioritise the well-being of employees, customers, and communities, fostering a positive social impact. For example, finance organisations that embrace sustainable practices often prioritise employee well-being by providing a healthy work environment, promoting work-life balance, and offering opportunities for professional development. This not only improves employee satisfaction but also enhances productivity and creativity.

Moreover, sustainable business models promote diversity and inclusion. Companies that adopt these models are committed to creating an inclusive workplace culture that values diversity in gender, race, ethnicity, and background. This leads to a more diverse and inclusive finance industry, where different perspectives and ideas are valued, resulting in better decision-making and innovation.

Sustainable business models also prioritise customer well-being. By offering sustainable products and services, finance organisations can meet the growing consumer demand for environmentally friendly solutions. This not only attracts environmentally conscious customers but also enhances customer loyalty and satisfaction. Furthermore, sustainable business models often involve engaging with local communities and supporting social causes, thereby building strong relationships and contributing to community development.

While sustainability is often associated with altruism, sustainable business models also bring significant *financial benefits* to organisations. By adopting sustainable practices, finance organisations can reduce operational costs and increase operational efficiency. For instance, energy-efficient buildings and equipment can lead to lower utility bills, while streamlined supply chains can minimise transportation costs. Furthermore, sustainable business models often result in improved risk management by identifying and addressing potential environmental and social risks that could impact the organisation's financial performance.

Sustainable business models also open up new market opportunities. As consumers become more environmentally conscious, there is a growing demand for sustainable financial products and services. By offering innovative and sustainable solutions, finance organisations can tap into this market and gain a competitive advantage. Additionally, sustainable practices can enhance brand reputation and attract socially responsible investors, leading to increased financial support and funding opportunities.

The adoption of sustainable business models is transforming the finance industry in various ways. Firstly, it is driving a shift towards responsible investing. Sustainable finance, also known as environmental, social, and governance (ESG) investing, has gained significant traction in recent years. Investors are increasingly considering environmental and social factors when making investment decisions, encouraging finance organisations to integrate sustainability into their investment strategies. This shift towards responsible investing is reshaping the finance industry and influencing capital allocation decisions.

Secondly, sustainable business models are driving innovation in financial products and services. Finance organisations are developing new products and services that align with sustainable principles, such as green bonds, sustainable investment funds, and impact investing. These innovative solutions not only cater to the growing demand for sustainable financial products but also contribute to the development of a more sustainable economy.

Furthermore, sustainable business models are fostering collaboration and partnerships within the finance industry. Organisations are coming together to share best practices, collaborate on sustainability initiatives, and drive industry-wide change. This collaboration is essential in addressing global sustainability challenges and accelerating the transition towards a more sustainable future.

Several finance organisations have successfully implemented sustainable business models, setting an example for others to follow. One such example is *Triodos Bank*, a European bank that only lends to organisations that have a positive social, environmental, or cultural impact. Triodos Bank has demonstrated that it is possible to be financially successful while prioritising sustainability.

Another example is *BlackRock*, the world's largest asset manager, which has integrated sustainability into its investment approach. BlackRock considers ESG factors in its investment decisions and actively engages with companies to drive positive change. By doing so, BlackRock has shown that sustainable investing can deliver long-term financial performance.

If you are considering implementing sustainable business models in your finance organisation, there are several steps you can take. Firstly, assess your organisation's current environmental and social impact and identify areas for improvement. This could involve conducting an environmental audit, engaging with stakeholders, and setting sustainability goals.

Secondly, develop a sustainability strategy that aligns with your organisation's mission and values. This strategy should include specific actions and targets to guide your organisation's sustainability efforts. It is important to involve employees at all levels of the organisation in the development and implementation of the strategy to ensure buy-in and commitment.

Next, integrate sustainability into your organisation's core operations. This could involve adopting energy-efficient practices, promoting recycling and waste reduction, and implementing responsible supply chain management. It is also important to train employees on sustainable practices and provide them with the necessary resources and support to implement these practices.

Additionally, engage with stakeholders, such as customers, suppliers, and investors, to communicate your organisation's commitment to sustainability and seek their input and collaboration. Building strong relationships with stakeholders is crucial in driving the adoption of sustainable business models.

While the benefits of sustainable business models are clear, there are also challenges and considerations to be aware of. One challenge is the need for financial resources to implement sustainable practices. It may require upfront investments to adopt energy-efficient technologies, implement recycling programmes, or develop sustainable products. However, these investments can lead to long-term cost savings and financial benefits.

Another challenge is the need for organisational change and employee engagement. Adopting sustainable business models often requires a shift in mindset and a change in organisational culture. It is important to communicate the benefits of sustainability to employees and involve them in the decision-making process. This can be achieved through training, awareness campaigns, and incentives.

Furthermore, measuring and reporting sustainability performance can be challenging. It is important to develop robust metrics and reporting frameworks to track progress and communicate the impact of sustainable practices. Engaging with external stakeholders, such as sustainability rating agencies and industry associations, can provide guidance and support in this regard.

Sustainable business models are transforming the finance industry by *driving profitability and purpose*. These models integrate environmental, social, and financial considerations into the core operations of organisations, resulting in a positive impact on the environment, society, and financial performance. By adopting sustainable practices, finance organisations can reduce their environmental footprint, enhance employee well-being, meet customer demand for sustainable solutions, and gain a competitive advantage.

The rise of sustainable business models is reshaping the finance industry, leading to responsible investing, innovative financial products, and collaboration among organisations. While there are challenges to overcome, the benefits of sustainable business models far outweigh the costs. By implementing sustainable business models, finance organisations can contribute to a more sustainable future while driving profitability and purpose.

CHAPTER 23

ESG Transformation

Addressing Climate Change, Biodiversity, and Social Impact

IN TODAY'S WORLD, BUSINESSES are increasingly recognising the importance of environmental, social, and governance (ESG) factors in their operations. ESG encompasses a wide range of issues, including the impact of business activities on the environment, the treatment of employees and communities, and the effectiveness of corporate governance. By integrating these factors into their decision-making processes, companies can not only mitigate risks but also drive sustainable growth.

Su et al. (2023) contend it is proven that companies can improve their ESG performance through digital transformation. Empirical testing reveals that digital transformation indeed has a positive impact on enterprises' ESG performance, and digital technology innovation can enhance ESG performance through dynamic capabilities such as green innovation, social responsibility, and operational management (Su et al., 2023).

When it comes to the environmental aspect of ESG, businesses need to consider the impact of their operations on the planet. This includes reducing greenhouse gas emissions, conserving natural resources, and minimising waste generation. By adopting sustainable practices, such as using renewable energy sources, implementing efficient waste management systems, and promoting recycling, companies can significantly reduce their environmental footprint.

The social aspect of ESG focuses on how businesses interact with their employees, customers, and communities. It includes factors such as fair labour practices, diversity and inclusion, and community engagement. By treating employees with respect, providing fair wages and benefits, and fostering a diverse and inclusive work environment, companies can enhance their reputation and attract and retain top talent.

Finally, the governance aspect of ESG pertains to the way companies are managed and governed. It includes factors such as board composition, executive compensation, and transparency in financial reporting. By implementing strong corporate governance

DOI: 10.1201/9781003514503-23

practices, companies can enhance accountability, reduce the risk of fraud and corruption, and build trust with their stakeholders.

ESG factors are increasingly gaining importance in finance transformation initiatives. Traditionally, financial decision-making focused primarily on maximising shareholder value without necessarily considering the broader impact of business activities. However, this narrow approach is no longer sufficient in a world facing pressing environmental and social challenges.

ESG finance transformation involves integrating ESG factors into financial decision-making processes. It requires businesses to assess the potential environmental and social risks and opportunities associated with their investments and to align their financial strategies with sustainability goals. By doing so, companies can better manage risks, identify new growth opportunities, and enhance long-term value creation.

Moreover, ESG finance transformation is essential for businesses to stay ahead of evolving regulatory requirements and investor expectations. Governments around the world are increasingly introducing regulations that require companies to disclose their ESG performance. Investors are also becoming more interested in investing in businesses that demonstrate strong ESG practices. By embracing ESG finance transformation, companies can ensure compliance with regulations, attract a wider pool of investors, and enhance their reputation in the market.

Fu (2023) argues ESG positively and significantly affects corporate financial performance, and digital transformation is the key driver of this effect (Fu and Li, 2023).

ESG finance transformation is a powerful driver of *sustainability*. By integrating ESG factors into financial decision-making, businesses can align their strategies with sustainability goals and contribute to the transition to a more sustainable economy.

One way ESG finance transformation drives sustainability is by encouraging businesses to invest in environmentally friendly technologies and practices. For example, companies may choose to invest in renewable energy projects, energy-efficient infrastructure, or sustainable agriculture. These investments not only reduce environmental impacts but also create new business opportunities and stimulate economic growth.

ESG finance transformation also promotes responsible investment practices. Investors are increasingly considering ESG factors when making investment decisions. By integrating ESG considerations into their investment analysis, investors can identify companies that are well-positioned to manage emerging risks and capitalise on sustainability opportunities. This, in turn, encourages businesses to improve their ESG performance, as it becomes a key determinant of their access to capital.

Furthermore, ESG finance transformation fosters transparency and accountability. By disclosing their ESG performance, businesses can provide stakeholders with valuable information about their environmental and social impacts. This transparency enhances trust and allows stakeholders to make informed decisions, whether they are employees, customers, or investors.

ESG finance transformation offers several benefits for businesses and investors alike. For businesses, adopting ESG finance transformation can lead to improved risk management, enhanced brand reputation, and increased access to capital.

By integrating ESG factors into their decision-making processes, businesses can identify and mitigate potential risks. For example, by assessing the environmental risks associated with their operations, companies can implement measures to reduce their exposure to climate change-related risks, such as extreme weather events or resource scarcity. This proactive approach to risk management not only protects businesses from potential financial losses but also enhances their long-term resilience.

ESG finance transformation also enhances a company's brand reputation. In today's socially conscious world, consumers are increasingly demanding products and services from companies that demonstrate a commitment to sustainability. By adopting sustainable practices and showcasing their ESG performance, businesses can attract environmentally and socially conscious consumers, which can lead to increased sales and customer loyalty.

Moreover, ESG finance transformation can improve a company's access to capital. Investors are increasingly looking for businesses that demonstrate strong ESG practices. By integrating ESG factors into their financial strategies and disclosing their ESG performance, companies can attract a wider pool of investors, including those who prioritise sustainable investments.

This increased investor interest can provide businesses with additional capital and potentially lower their cost of capital.

For investors, ESG finance transformation offers the opportunity to align their investment portfolios with their values. By integrating ESG considerations into their investment analysis, investors can support companies that are making a positive impact on the environment and society. This alignment between investment decisions and personal values can provide investors with a sense of purpose and contribute to their overall financial well-being.

Numerous companies have already embraced the ESG finance transformation and are reaping the benefits. The following are a few case studies to help understand how ESG finance transformation can drive sustainability:

General Motors: General Motors, a global manufacturing company, implemented an ESG finance transformation by investing in renewable energy sources to power its operations. By transitioning to renewable energy, it is reducing its greenhouse gas emissions and significantly decreasing its reliance on fossil fuels. This not only helped the company reduce its environmental impact but also resulted in cost savings, as renewable energy sources became more cost-effective over time. Its supply chain boasts transparency and focused initiatives on workplace safety and environmentally-friendly innovation. It has a company-wide ethos of 'zero crashes, zero emissions, and zero congestion' and is viewed as a leader in automotive sustainability. The firm's directors regularly review the company's ESG performance and progress, ensuring it remains a central focus for global operations in line with shifting global expectations and regulations. GM recently announced an innovative new pledge, inviting global suppliers to join the company in a commitment to carbon neutrality, the development of social responsibility programmes, and the implementation of sustainable procurement practices in its supply chain operations[1].

FedEx: FedEx integrated ESG factors into its supply chain management practices. The company implemented a supplier code of conduct that focused on fair labour practices and environmental stewardship. By working closely with its suppliers and providing them with the necessary support and resources, FedEx improved the working conditions of its suppliers' employees and reduced the environmental impact of its supply chain. This approach not only enhanced the company's brand reputation but also resulted in increased supplier loyalty and improved supply chain resilience. As one of the world's leading logistics providers, FedEx's environmental responsibilities are a key aspect of the company's global supply chains.

Responsible environmental practices have been embedded across its operations, designed to boost efficiency while simultaneously cutting both waste and emissions, and it's working towards ambitious climate-oriented targets. Having worked to cut its aircraft emissions by 30% and increased its Express fleet's efficiency by 30% by 2020, FedEx's sights are now set on its 2030 goals when it's thought that 30% of the company's jet fuel will consist of alternative fuel sources.

HSBC: HSBC bank embraced ESG finance transformation by developing a comprehensive framework to assess the ESG performance of potential investment opportunities. HSBC has been one of the leaders amongst its peers in implementing ESG. By integrating ESG considerations into its investment analysis, HSBC was able to identify areas of investment that demonstrated strong ESG practices and were well-positioned to manage emerging risks. This approach allowed the company to generate attractive risk-adjusted returns for its investors while contributing to the transition to a more sustainable economy.

These case studies highlight the diverse ways in which companies can implement ESG finance transformation and drive sustainability. Whether it's through investing in renewable energy, improving supply chain practices, or integrating ESG considerations into investment decisions, businesses can make a positive impact on the environment and society while also enhancing their financial performance.

Implementing ESG finance transformation in your organisation requires a comprehensive and strategic approach. Here are some key steps to consider:

Assess Your Current ESG Performance: Start by evaluating your organisation's current ESG practices and performance. Identify areas where improvements can be made and set specific goals and targets for ESG integration.

Engage Stakeholders: ESG finance transformation requires the support and involvement of various stakeholders, including employees, customers, investors, and regulators.

Engage with these stakeholders and communicate the importance of ESG integration and the benefits it can bring.

Integrate ESG into Decision-Making Processes: Develop frameworks and tools to integrate ESG factors into your organisation's decision-making processes. This may involve revising existing financial models, developing new risk assessment methodologies, and enhancing reporting and disclosure practices.

Train and Educate Employees: ESG finance transformation requires a shift in mindset and behaviour. Provide training and education to employees to ensure they understand the importance of ESG and how it can be integrated into their daily work.

Monitor and Report Progress: Regularly monitor and report on your organisation's progress in implementing ESG finance transformation. This will help you track your performance, identify areas for improvement, and communicate your ESG achievements to stakeholders.

Remember, ESG finance transformation is a journey, and it requires ongoing commitment and continuous improvement. By taking the necessary steps to integrate ESG into your organisation's financial decision-making processes, you can drive sustainability and position your business for long-term success.

Implementing ESG finance transformation can be facilitated by leveraging various *tools and technologies.* These tools can help organisations collect, analyse, and report ESG data, as well as monitor and manage ESG risks and opportunities. Here are some examples:

ESG Data Management Platforms: These platforms allow organisations to collect, store, and analyse ESG data from various sources, such as financial reports, sustainability reports, and third-party databases. They provide tools for data validation, quality assurance, and integration, enabling organisations to generate reliable and consistent ESG insights.

ESG Analytics and Reporting Software: These software solutions help organisations analyse and report on their ESG performance. They provide tools for data visualisation, trend analysis, and benchmarking, allowing organisations to track their progress and compare their performance against industry peers.

Risk Management Systems: These systems enable organisations to identify and manage ESG risks. They provide tools for risk assessment, scenario analysis, and mitigation planning, helping organisations proactively address potential ESG risks and enhance their resilience.

Stakeholder Engagement Platforms: These platforms facilitate engagement with stakeholders, such as employees, customers, investors, and regulators. They provide tools for collecting feedback, conducting surveys, and managing stakeholder relationships, enabling organisations to understand stakeholder expectations and incorporate their input into decision-making processes.

Blockchain Technology: Blockchain technology can enhance the transparency and traceability of ESG data. By using blockchain, organisations can securely record and share ESG information, ensuring its integrity and immutability. This can help build trust among stakeholders and facilitate the verification of ESG claims.

These tools and technologies can support organisations in their ESG finance transformation journey by providing them with the necessary infrastructure and capabilities to effectively manage ESG risks and opportunities. However, it's important to note that technology alone is not sufficient. Organisations also need to develop the necessary skills and capabilities to use these tools effectively and interpret and act upon the insights they provide.

While ESG finance transformation offers numerous benefits, it also comes with its fair share of challenges and risks. Here are some key challenges to consider:

Data Availability and Quality: One of the main challenges in ESG finance transformation is the availability and quality of ESG data. ESG data can be fragmented, inconsistent, and difficult to compare across companies and industries. Organisations need to invest in robust data collection and management systems to ensure the reliability and accuracy of their ESG insights.

Measurement and Reporting Standards: Another challenge is the lack of standardised measurement and reporting frameworks for ESG performance. Organisations often struggle to define and measure ESG indicators in a consistent and comparable manner. The development of globally accepted standards and frameworks can help address this challenge and provide organisations with clear guidelines for ESG reporting.

Integration into Financial Models: Integrating ESG factors into financial models can be complex, as it requires a deep understanding of the interdependencies between financial and non-financial performance. Organisations need to develop sophisticated modelling techniques and tools to accurately assess the financial implications of ESG risks and opportunities.

Resistance to Change: ESG finance transformation requires a cultural shift within organisations. It may face resistance from employees who are not familiar with ESG concepts or who perceive them as conflicting with financial objectives. Organisations need to invest in change management initiatives and employee education to overcome this resistance and foster a culture of sustainability.

Greenwashing and Reputation Risks: As ESG becomes more prominent, there is a risk of greenwashing, where companies make false or misleading claims about their ESG performance. This can damage their reputation and lead to legal and regulatory consequences. Organisations need to ensure the accuracy and transparency of their ESG disclosures and implement robust governance mechanisms to prevent greenwashing.

It's important for organisations to be aware of these challenges and risks and develop strategies to address them. By proactively managing these challenges, organisations can maximise the benefits of ESG finance transformation and effectively drive sustainability.

ESG finance transformation is not just a passing trend; it is here to stay. As the environmental and social challenges facing the world continue to intensify, ESG will become even more important in financial decision-making.

In the future, we can expect to see increased regulatory requirements around ESG reporting and disclosure. Governments and regulatory bodies are recognising the importance of ESG factors in driving sustainability and are taking steps to ensure transparency and accountability. Organisations will need to stay up to date with evolving regulations and adapt their ESG practices accordingly.

Moreover, we can expect to see further integration of ESG factors into investment analysis and decision-making processes. Investors are increasingly demanding ESG data and insights to support their investment decisions. This trend will continue to drive the demand for robust ESG analytics tools and platforms.

Furthermore, technology will play a critical role in advancing ESG finance transformation. Innovations such as artificial intelligence, machine learning, and big data analytics will enable organisations to collect, analyse, and report on ESG data more effectively. Blockchain technology will enhance the transparency and traceability of ESG information, further building trust among stakeholders.

Ultimately, the future of ESG finance transformation lies in the hands of businesses, investors, and regulators. By embracing ESG and integrating it into their decision-making processes, organisations can drive sustainability, create long-term value, and contribute to a more sustainable and inclusive economy.

ESG finance transformation is a powerful driver of sustainability. By integrating ESG factors into financial decision-making, businesses can align their strategies with sustainability goals and contribute to the transition to a more sustainable economy. ESG finance transformation offers numerous benefits for businesses, including improved risk management and enhanced brand reputation.

The Impact of Climate Change on Nature and Biodiversity: Climate change has emerged as one of the most pressing challenges of our time, with far-reaching effects on the environment and ecosystems. One of the areas that has been significantly impacted by climate change is nature and biodiversity. Rising temperatures, changing rainfall patterns, and extreme weather events have disrupted ecosystems and threatened the survival of many species. The loss of biodiversity not only affects the intricate balance of ecosystems but also has implications for human well-being. It is therefore crucial to understand the link between climate change and nature and the role that finance transformation can play in addressing this issue.

Understanding the Role of Finance Transformation in Addressing Climate Change: Finance transformation means redefining and reorganising financial systems and practices to meet the challenges of the modern world. In the context of climate change, finance transformation involves aligning financial strategies and investments with sustainability goals. By integrating environmental considerations into financial decision-making, organisations can contribute to the efforts of mitigating climate change and protecting nature. Finance transformation involves not only the adoption of sustainable finance practices but also the development of innovative financial instruments and strategies to support nature conservation.

The Importance of Sustainable Finance In Protecting Nature: Sustainable finance is a key component of finance transformation and plays a crucial role in protecting nature. It involves the integration of ESG factors into financial decision-making processes. By considering the environmental impact of investments and directing capital towards environmentally friendly projects, sustainable finance can help drive the transition to a low-carbon and nature-friendly economy. Sustainable finance also encourages transparency and accountability, ensuring that financial institutions and investors are held responsible for the environmental consequences of their actions. This shift towards sustainable finance is essential for safeguarding nature and biodiversity in the face of climate change.

Key Initiatives and Strategies: To achieve finance transformation for climate change and nature conservation, several key initiatives and strategies need to be implemented. One such initiative is the development of green bonds, which are financial instruments specifically designed to fund environmentally friendly projects. These bonds attract investors who are looking to support sustainable initiatives while also providing them with a financial return. Another strategy is the integration of ESG factors into investment decision-making, which involves assessing the environmental and social impact of potential investments. This approach helps direct capital towards nature-friendly projects and encourages companies to adopt sustainable practices.

Numerous successful finance transformation projects have already had a significant impact on climate change and nature conservation. One such example is the *Global Environment Facility (GEF)*, which provides grants to projects that address climate change, biodiversity loss, and land degradation. The GEF has supported initiatives in various countries, ranging from reforestation projects to the development of renewable energy sources. Another case study is the *Sustainable Forestry Initiative (SFI)*, which promotes responsible forest management practices. By working with stakeholders across the supply chain, SFI ensures that forests are managed in a way that protects biodiversity and mitigates climate change.

Finance transformation for sustainability cannot be achieved without the active involvement of various stakeholders. Governments play a crucial role in creating an enabling policy environment that encourages sustainable finance practices. They can introduce regulations and incentives that promote green investments and hold financial institutions accountable for their environmental impact. Financial institutions themselves need to adopt sustainable finance practices and integrate ESG factors into their decision-making processes. Investors also have a role to play by directing their capital towards sustainable investments and demanding transparency from financial institutions. Collaboration between all these stakeholders is essential for driving finance transformation and ensuring a sustainable future for nature and biodiversity.

While finance transformation holds great promise for addressing climate change and protecting nature, it is not without its challenges and barriers. One of the major hurdles is the lack of awareness and understanding of sustainable finance practices among financial institutions and investors. Many organisations are still focused on short-term financial gains and are not fully aware of the long-term risks associated with climate change. Another challenge is the complexity of integrating environmental considerations into financial decision-making processes. Assessing the environmental impact of investments requires reliable data and methodologies, which may not always be readily available. Overcoming these challenges requires education, capacity-building, and collaboration among stakeholders.

Finance transformation holds tremendous potential for addressing climate change and protecting biodiversity. As more organisations recognise the importance of sustainable finance, we can expect to see an increase in green investments and the development of innovative financial instruments. This shift towards sustainable finance will not only contribute to global efforts to mitigate climate change but also help safeguard nature and biodiversity. By aligning financial strategies with sustainability goals, we can create a more resilient and environmentally friendly economy that benefits both present and future generations.

Embarking on finance transformation for sustainability can be a complex process, but there are several resources and tools available to organisations that are looking to get started. The United Nations Environment Programme provides guidance and best practices for sustainable finance through various publications and reports. The Principles for Responsible Investment is another valuable resource that offers a framework for integrating ESG factors into investment decision-making. Additionally, organisations can collaborate with sustainability consultants and experts who can provide tailored advice and support throughout the finance transformation journey.

There is a crucial role of finance transformation in helping address the issues of climate change and nature conservation. Finance transformation offers a unique opportunity to align financial strategies with sustainability goals, thereby contributing to global efforts to mitigate climate change and protect biodiversity. Through sustainable finance practices, organisations can direct capital towards nature-friendly projects and support the transition to a low-carbon economy. However, achieving finance transformation for sustainability requires the active involvement of various stakeholders, including governments, financial institutions, and investors. By working together, we can harness the power of finance to make a real difference in the fight against climate change and the conservation of nature.

23.1 SOCIAL IMPACT

As the world becomes increasingly aware of the social and environmental challenges we face, the role of *social impact* in finance transformation has gained significant importance. Social impact is the positive change that an organisation brings to society through its activities, while finance transformation involves the strategic rethinking of financial processes and systems. It is becoming more important for CFOs to understand the connections between social impact and financial sustainability, as this is crucial for businesses and investors alike.

Financial sustainability is no longer solely about maximising profits. It now encompasses the broader goal of creating value for all stakeholders, including employees, communities, and the environment. By incorporating social impact into finance transformation, organisations can align their financial decisions with their values and contribute to a more sustainable future.

Social impact plays a pivotal role in financial sustainability. By considering the social and environmental consequences of their actions, organisations can mitigate risks and enhance their long-term financial performance. A growing body of research suggests that companies with strong social and environmental practices tend to outperform their peers in terms of profitability and shareholder value.

Moreover, integrating social impact into financial decision-making helps organisations build trust and enhance their reputation. In an era of increasing social awareness, consumers and investors are demanding greater transparency and accountability from businesses. By demonstrating a commitment to social impact, organisations can attract and retain customers, investors, and talented employees.

Social impact investing has emerged as a powerful tool for driving social change while generating financial returns. This approach involves strategically deploying capital to achieve both financial and social objectives. Several key principles guide social impact investing:

Intentionality: Social impact investors are purpose-driven, seeking to create positive change through their investments. They actively seek out opportunities that align with their social and environmental goals.

Measurability: Measuring social impact is essential for effective decision-making and accountability. Social impact investors employ various metrics and frameworks to assess the outcomes and effectiveness of their investments.

Additionality: Social impact investing aims to go beyond traditional philanthropy by creating sustainable, scalable solutions to social and environmental challenges. It seeks to generate both financial returns and positive social outcomes.

By adhering to these principles, social impact investors can drive meaningful change while generating financial returns.

Measuring social impact is a complex task, but it is essential for understanding the effectiveness of social impact initiatives and making informed financial decisions. Traditional financial metrics such as return on investment do not capture the full picture of social impact. Instead, organisations need to employ a range of tools and methodologies to assess their social and environmental performance.

One widely used approach is the Social Return on Investment (SROI) framework. SROI quantifies the social, environmental, and economic value created by an organisation and compares it to the resources invested. By assigning a monetary value to social impact, SROI enables organisations to evaluate the effectiveness of their initiatives and make data-driven decisions.

Additionally, organisations can leverage other tools, such as impact assessments, stakeholder engagement, and sustainability reporting, to measure and communicate their social impact.

Integrating social impact into financial decision-making requires a shift in mindset and a re-evaluation of traditional financial models. Organisations need to consider social and environmental factors alongside financial considerations when making investment decisions.

One approach is impact investing, which involves deploying capital specifically to generate positive social and environmental impact. Impact investors actively seek out investment opportunities that align with their values and contribute to sustainable development.

Another approach is the incorporation of ESG criteria into investment analysis. ESG factors consider an organisation's performance in areas such as climate change, labour practices, and board diversity. By integrating ESG criteria into financial analysis, organisations can assess the long-term sustainability and resilience of their investments.

To illustrate the power of social impact in finance transformation, let's examine two case studies showcasing successful initiatives:

Microsoft: Microsoft, a global technology firm, decided to invest a portion of its profits in renewable energy projects. By aligning its financial resources with its commitment to sustainability, Microsoft not only reduced its carbon footprint but also generated additional revenue streams from renewable energy sales.

Morgan Stanley: Morgan Stanley adopted a social impact investing strategy, focusing on investments that promote education in underserved communities. By supporting educational initiatives, the fund not only generated financial returns for its investors but also contributed to the development of future generations.

These case studies demonstrate how social impact can drive positive change while delivering financial benefits.

While incorporating social impact into financial sustainability is a compelling proposition, it also comes with challenges and limitations. Some of the key hurdles include:

Measuring Intangible Impact: Quantifying social impact can be challenging, especially when it comes to intangible outcomes such as improved quality of life or enhanced social cohesion. Organisations need to develop robust methodologies to capture and evaluate these intangible impacts.

Balancing Financial and Social Objectives: Balancing financial returns with social impact goals can be a delicate task. Organisations must strike a balance between generating profits and creating positive change, which may require trade-offs and careful decision-making.

Limited Data Availability: Access to reliable data on social impact can be limited, making it difficult for organisations to assess the effectiveness of their initiatives. Collaboration between organisations, governments, and non-profits is essential to improve data availability and quality.

Despite these challenges, organisations can overcome them by adopting a strategic and long-term approach to social impact.

To effectively implement social impact initiatives in finance, organisations can follow several strategies:

Set Clear Social Impact Goals: Organisations should define specific social impact goals that align with their mission and values. These goals should be measurable, time-bound, and aligned with stakeholders' expectations.

Integrate Social Impact into Strategic Planning: Social impact should be integrated into the organisation's strategic planning processes. It should be considered alongside financial objectives to ensure that both are pursued in a balanced manner.

Engage Stakeholders: Engaging stakeholders, including employees, customers, and communities, is crucial for the success of social impact initiatives. By involving stakeholders in the decision-making process, organisations can gain valuable insights and build support for their initiatives.

Collaborate with Partners: Collaboration with external partners, such as non-profits, governments, and other businesses, can amplify the impact of social impact initiatives. By leveraging complementary expertise and resources, organisations can achieve greater social and financial outcomes.

Several tools and resources are available to help organisations measure and manage their social impact:

Global Impact Investing Network (GIIN): The GIIN provides resources, research, and best practices for impact investors. It offers guidance on impact measurement and management, helping organisations assess the social and environmental performance of their investments.

B Impact Assessment: The B Impact Assessment is a comprehensive tool that measures a company's impact on its workers, community, environment, and governance. It provides a holistic view of a company's social and environmental performance and helps identify areas for improvement.

The United Nations Sustainable Development Goals (SDGs): The SDGs provide a globally recognised framework for organisations to align their social impact initiatives with broader global goals. By aligning with the SDGs, organisations can contribute to international efforts to address social and environmental challenges.

As we navigate an increasingly complex and interconnected world, the role of social impact in financial sustainability will continue to grow in importance. Organisations that embrace social impact as a core value and integrate it into their financial decision-making will be better positioned to create value for all stakeholders and contribute to a more sustainable future.

By understanding the principles of social impact investing, measuring social impact, and effectively implementing social impact initiatives, organisations can drive positive change while achieving financial sustainability. While challenges and limitations exist, organisations can overcome them through strategic planning, stakeholder engagement, and collaboration.

The future of finance lies in the hands of those who recognise the power of social impact and embrace it as a catalyst for meaningful change. Organisations are being increasingly scrutinised, so it is important we display leadership to make a difference and create a more sustainable and equitable world.

NOTE

1. https://supplychaindigital.com/sustainability/top-10-esg-friendly-supply-chains

CHAPTER 24

Organisation Design

IN TODAY'S RAPIDLY CHANGING business landscape, finance transformation has become a key priority for organisations seeking to unlock financial success. Finance transformation refers to the process of reimagining and reinventing the finance function within an organisation to drive efficiency, effectiveness, and value creation. It involves the integration of people, processes, and technology to streamline financial operations and enable strategic decision-making. At the heart of finance transformation lies the concept of organisational design, which focuses on aligning the structure, roles, and responsibilities of an organisation to achieve its strategic objectives. By combining finance transformation and organisational design, organisations can create a powerful framework that drives sustainable growth and competitive advantage.

Kapotas (2023) argues digital transformation within finance can help in organisational design through streamlining, optimising resource allocation, and driving culture change (Kapotas, 2023). Digital transformation is impacting organisation design in many dimensions, but the change already started decades ago. Khashabi et al. (2019) argues digitisation will continue to impact organisation design and will accelerate in the near future, since transformation is starting to reach more applications, domains, and locations (Khashabi and Kretschmer, 2019).

In this fast-paced and dynamic business environment, organisations face numerous challenges that impact their financial performance. These challenges include increased competition, evolving customer expectations, technological advancements, and regulatory changes. To navigate these challenges successfully, organisations need to have a finance function that is agile, data-driven, and proactive. This is where finance transformation comes into play. By transforming the finance function, organisations can enhance their ability to adapt to change, make informed decisions, and drive value creation. Finance transformation enables organisations to streamline financial processes, improve operational efficiency, and optimise resource allocation. It also empowers finance professionals to become strategic business partners, providing valuable insights and guidance to support organisational growth.

 DOI: 10.1201/9781003514503-24

Finance transformation encompasses a wide range of activities, each of which plays a crucial role in enhancing the effectiveness and efficiency of the finance function. Some key elements of finance transformation include:

Process Optimisation: Finance transformation involves reviewing and reengineering financial processes to eliminate inefficiencies, reduce costs, and improve the speed and accuracy of financial reporting. This includes automating manual tasks, implementing standardised procedures, and leveraging technology to streamline workflows.

Data Analytics and Insights: In the era of big data, finance transformation emphasises the importance of leveraging data analytics to drive business insights. By harnessing the power of data, organisations can gain a deeper understanding of their financial performance, identify trends and patterns, and make data-driven decisions.

Talent Development: Finance transformation recognises the critical role of talent in driving organisational success. It involves investing in the development of finance professionals and equipping them with the necessary skills and knowledge to excel in their roles. This includes providing training and development opportunities, fostering a culture of continuous learning, and attracting top finance talent.

Technology Enablement: Technology plays a pivotal role in finance transformation. By leveraging advanced financial systems, automation tools, and analytics platforms, organisations can enhance the efficiency, accuracy, and reliability of financial processes. Technology also enables real-time reporting, data visualisation, and scenario planning, empowering finance professionals to make informed decisions.

The integration of finance transformation and organisational design can have numerous benefits for organisations. By aligning the finance function with the strategic objectives of the organisation, organisations can achieve the following benefits:

Enhanced Agility and Adaptability: Finance transformation enables organisations to respond quickly and effectively to changes in the business environment. By streamlining processes, leveraging technology, and fostering a culture of innovation, organisations can become more agile and adaptable, allowing them to seize new opportunities and mitigate risks.

Improved Decision-Making: By leveraging data analytics and insights, finance transformation enables organisations to make informed, data-driven decisions. This leads to improved financial forecasting, better resource allocation, and enhanced risk management. With accurate and timely financial information, decision-makers can confidently navigate the complexities of the business landscape.

Increased Operational Efficiency: Finance transformation involves optimising financial processes, automating manual tasks, and eliminating redundancies. This leads to increased operational efficiency, reduced costs, and improved productivity. By eliminating time-consuming manual tasks, finance professionals can focus on value-added activities such as financial analysis and strategic planning.

Strengthened Governance and Compliance: Finance transformation emphasises the importance of strong governance and compliance within the finance function. By implementing robust controls, standardised procedures, and automated reporting, organisations can enhance their compliance with regulatory requirements and internal policies. This leads to improved risk management and increased stakeholder confidence.

Strategic Partnership: Finance transformation enables finance professionals to become strategic partners in the business. By providing valuable insights, conducting financial analysis, and contributing to strategic decision-making, finance professionals can add value beyond traditional financial reporting. This positions finance as a trusted advisor and strategic enabler within the organisation.

To illustrate the power of finance transformation and organisational design, let's explore two successful case studies:

Honda: Honda, a multinational manufacturing company, embarked on a finance transformation journey to enhance its financial performance and competitiveness. Through process optimisation and automation, the company streamlined its financial reporting and reduced the time required for month-end closing. This resulted in faster decision-making and improved cash flow management. Additionally, by investing in talent development and technology enablement, Honda transformed its finance function into a strategic partner for the business. Finance professionals were empowered to provide valuable insights and guidance, contributing to the company's growth strategy.

MoonPay: MoonPay, a technology startup, recognised the need for finance transformation to support its rapid growth and scalability. By implementing cloud-based financial systems and analytics platforms, the company gained real-time visibility into its financial performance.

This enabled the finance team to make data-driven decisions and optimise resource allocation. Additionally, MoonPay redesigned its finance organisation structure to align with its strategic priorities. This involved creating cross-functional teams and introducing new roles to support business expansion. As a result, MoonPay achieved enhanced operational efficiency, improved financial controls, and a strong foundation for future growth.

Implementing finance transformation in your organisation requires careful planning and execution. Here are some key steps to consider:

Assess the Current State: Evaluate the current state of your finance function, including processes, technology, talent, and organisational structure. Identify areas for improvement and set clear objectives for finance transformation.

Define the Future State: Envision the desired future state of your finance function. Define the goals, outcomes, and benefits you aim to achieve through finance transformation. This will serve as a roadmap for your transformation journey.

Develop a Transformation Plan: Create a detailed plan that outlines the activities, timelines, and resources required for finance transformation. Assign responsibilities and establish clear communication channels to ensure smooth execution.

Engage Stakeholders: Engage key stakeholders, including finance leaders, business leaders, and employees, in the finance transformation journey. Seek their input, address concerns, and build alignment around the vision and objectives of the transformation.

Implement Changes: Execute the transformation plan, implementing changes in processes, technology, talent development, and organisational design. Monitor progress closely, making adjustments as necessary to ensure successful implementation.

Measure and Monitor: Establish key performance indicators (KPIs) to measure the effectiveness and impact of finance transformation. Regularly monitor and assess the outcomes against the defined objectives. Use the insights gained to drive continuous improvement.

While finance transformation offers significant benefits, it is not without its challenges. Some common challenges and obstacles include:

Resistance to Change: Finance transformation often involves changes to established processes, roles, and responsibilities. Resistance to change can arise from employees who are comfortable with the status quo or fear the unknown. To overcome this challenge, organisations should communicate the need for change, provide training and support, and involve employees in the transformation process.

Lack of Alignment: Finance transformation requires alignment between finance, IT, and other business functions. Lack of alignment can hinder the implementation of new technologies, the integration of systems, and collaboration between teams. To address this challenge, organisations should foster a culture of collaboration, establish clear communication channels, and involve stakeholders in decision-making processes.

Data Quality and Availability: Effective finance transformation relies on accurate and reliable data. However, many organisations struggle with data quality issues, including data inconsistencies, duplication, and lack of integration. To overcome this challenge, organisations should invest in data governance, establish data quality standards, and leverage technology solutions for data integration and validation.

Resource Constraints: Implementing finance transformation requires financial and human resources. Limited budgets and talent shortages can pose challenges to organisations. To address this, organisations should prioritise transformation initiatives, seek external partnerships or outsourcing options, and leverage technology to automate manual tasks and free up resources.

To maximise the success of finance transformation and organisational design, organisations should consider the following best practices:

Align with Strategic Objectives: Ensure that finance transformation initiatives are aligned with the overall strategic objectives of the organisation. This will help prioritise initiatives, allocate resources effectively, and measure the impact of transformation on organisational performance.

Foster a Culture of Innovation: Encourage a culture of innovation and continuous improvement within the finance function. Empower finance professionals to challenge the status quo, experiment with new ideas, and embrace emerging technologies. This will drive innovation and enable the finance function to stay ahead of the curve.

Invest in Talent Development: Recognise the importance of talent in driving finance transformation. Invest in the development of finance professionals, providing them with the necessary skills and knowledge to excel in their roles. This includes offering training and mentoring programmes, encouraging cross-functional collaboration, and promoting a culture of learning.

Leverage Technology Strategically: Technology plays a pivotal role in finance transformation. However, it is important to leverage technology strategically, focusing on solutions that align with the organisation's needs and objectives. Conduct thorough research, engage with technology vendors, and ensure seamless integration with existing systems.

Continuously Monitor and Evaluate: Monitor the effectiveness and impact of finance transformation initiatives on an ongoing basis. Establish key performance indicators (KPIs) to measure progress and regularly evaluate the outcomes against the defined objectives. Use the insights gained to drive continuous improvement and refine the finance transformation strategy.

Technology is a key enabler of finance transformation. It empowers organisations to automate manual tasks, enhance data accuracy, and gain real-time visibility into financial performance. Some key technologies that play a role in finance transformation include:

Cloud-Based Financial Systems: Cloud-based financial systems offer numerous benefits, including scalability, flexibility, and cost-effectiveness. They enable organisations to centralise financial data, streamline reporting processes, and improve collaboration between finance and other business functions.

Robotic Process Automation (RPA): RPA involves the use of software robots to automate repetitive and rule-based tasks. It can be applied to various finance processes, such as accounts payable, accounts receivable, and financial reconciliation. RPA improves accuracy, reduces processing time, and frees up finance professionals to focus on value-added activities.

Data Analytics and Visualisation Tools: Data analytics and visualisation tools enable organisations to derive insights from large volumes of financial data. They provide interactive dashboards, data visualisation, and predictive analytics capabilities, allowing finance professionals to gain a deeper understanding of financial performance and make informed decisions.

Artificial Intelligence (AI) and Machine Learning (ML): AI and ML technologies have the potential to revolutionise finance transformation. They can automate complex tasks, identify patterns and trends, and provide predictive analytics capabilities. AI-powered chatbots and virtual assistants can also enhance the finance user experience by providing real-time support and answering queries.

Finance transformation is no longer a luxury but a necessity for organisations seeking to unlock financial success. In an increasingly complex and competitive business landscape, finance transformation enables organisations to become more agile, data-driven, and strategically aligned. By integrating finance transformation with organisational design, organisations can create a powerful framework that drives sustainable growth and competitive advantage. The future of finance transformation lies in embracing emerging technologies, fostering a culture of innovation, and investing in talent development. As organisations continue to evolve and adapt, finance transformation will play a pivotal role in shaping the future of organisational design and driving financial success.

CHAPTER 25

Organisational Culture

ORGANISATIONAL CULTURE PLAYS A pivotal role in shaping the success of finance transformations within companies. It refers to the shared values, beliefs, and behaviours that define how people within an organisation think, act, and interact. Understanding the essence of organisational culture is crucial in comprehending its impact on finance transformations.

Culture is deeply ingrained in every aspect of an organisation, including its finance department. It influences the way financial decisions are made, the level of collaboration among team members, and the overall attitude towards change. Therefore, when embarking on a finance transformation journey, it is vital to assess and comprehend the existing culture within the company.

The influence of organisational culture on finance transformations cannot be overstated. A culture that embraces change, innovation, and continuous improvement creates an environment where finance transformations can thrive. On the other hand, a culture that resists change, favours traditional approaches, and lacks collaboration can hinder the success of finance transformations.

In a positive culture, employees are more open to adopting new technologies and methodologies, such as automation, analytics, and cloud-based solutions. They are willing to challenge the status quo, seek process improvements, and adapt to evolving market conditions. Such a culture fosters a dynamic and agile finance function capable of driving successful finance transformations.

A culture that supports successful finance transformations exhibits several key characteristics. Firstly, it encourages open communication and information sharing. This transparency allows for a free flow of ideas, feedback, and knowledge exchange, enabling the finance team to stay abreast of industry trends and best practices.

Secondly, a supportive culture values continuous learning and professional development. It promotes a growth mindset, where individuals are encouraged to acquire new skills and expand their knowledge base. This mindset is crucial for finance professionals as they navigate the changing landscape of finance transformations.

DOI: 10.1201/9781003514503-25

Lastly, a culture that supports successful finance transformations is one that values collaboration and teamwork. It fosters an environment where individuals work together towards a common goal, leveraging each other's strengths and expertise. This collaborative approach enhances problem-solving, decision-making, and overall effectiveness in executing finance transformations.

Developing a culture that drives successful finance transformations requires a strategic approach. Firstly, it is essential to clearly define the vision and goals of the finance transformation initiative. Communicating this vision to the entire organisation creates a sense of purpose and aligns everyone towards a common objective.

Secondly, leaders need to lead by example. They should embody the desired culture and behaviour, demonstrating the values and mindset necessary for successful finance transformations. This includes being open to change, encouraging innovation, and fostering a culture of continuous improvement.

Furthermore, organisations can establish forums and platforms for knowledge sharing and collaboration. This can be done through regular team meetings, cross-functional workshops, or the use of collaborative technology tools. These initiatives promote a culture of learning, communication, and collective problem-solving, which are essential for driving successful finance transformations.

Several companies have successfully harnessed the power of organisational culture to drive successful finance transformations. One such example is *Volkswagen Group*, a multinational corporation that embarked on a finance transformation journey to streamline its financial processes and improve decision-making.

Volkswagen Group recognised the importance of creating a culture that supports change and innovation. They encouraged their finance team to think outside the box, experiment with new technologies, and challenge traditional practices. They also established cross-functional teams to promote collaboration and foster a culture of continuous improvement.

As a result, Volkswagen Group witnessed significant improvements in their finance function. The finance team became more agile and efficient, leveraging automation and advanced analytics to drive data-driven decision-making. The positive culture they cultivated played a crucial role in their successful finance transformation.

Changing organisational culture is not without its challenges, especially during finance transformations. Resistance to change, fear of the unknown, and ingrained habits can hinder the process of cultural change. However, there are strategies that can help overcome these challenges.

Firstly, *communication* is key. Leaders should take the time to clearly communicate the reasons behind the finance transformation, its benefits, and the expected impact on individuals and the organisation as a whole. This helps alleviate fears and creates a sense of understanding and buy-in.

Secondly, *involving employees* in the change process can help overcome resistance. By including them in decision-making, seeking their input, and empowering them to be part of the solution, employees feel a sense of ownership and are more likely to embrace the cultural changes required for successful finance transformations.

Lastly, providing adequate *training and support* is crucial. Organisations should invest in training programmes that equip employees with the necessary skills and knowledge to adapt to the changing finance landscape. This not only boosts their confidence but also ensures they have the tools to succeed in the transformed finance function.

Leadership plays a vital role in shaping organisational culture for successful finance transformations. Leaders must set the tone from the top, embodying the desired culture and behaviour. They should be champions of change, inspiring and motivating their teams to embrace the finance transformation journey.

Leaders should actively communicate the vision, goals, and benefits of the finance transformation to the entire organisation. They should create a safe environment where employees feel comfortable challenging the status quo, experimenting with new ideas, and taking calculated risks. This encourages a culture of innovation and continuous improvement.

Furthermore, leaders should lead by example and demonstrate the desired values and behaviours. They should be accessible to their teams, actively listen to their concerns, and provide support when needed. By doing so, leaders foster a culture of trust, collaboration, and open communication, which are essential for successful finance transformations.

Measuring the effectiveness of organisational culture in driving finance transformations is crucial to ensure continuous improvement. There are several key performance indicators (KPIs) that can be used to assess the impact of culture on finance transformations.

Firstly, employee engagement surveys can provide insights into the level of employee satisfaction, motivation, and alignment with the desired culture. High levels of engagement indicate a positive culture that supports successful finance transformations.

Secondly, the adoption of new technologies and methodologies can be measured. Tracking the implementation and utilisation of automation tools, data analytics platforms, and cloud-based solutions can indicate the level of cultural readiness for finance transformations.

Lastly, the overall performance of the finance function can serve as a measure of the effectiveness of organisational culture. Key metrics such as process efficiency, accuracy of financial reporting, and timeliness of decision-making can reflect the impact of culture on finance transformations.

Organisational culture plays a pivotal role in driving successful finance transformations. A culture that embraces change, innovation, collaboration, and continuous learning creates an environment where finance transformations can thrive. Leaders must take a strategic approach to develop and nurture a culture that supports the goals and vision of the finance transformation initiative.

By understanding the impact of culture on finance transformations, organisations can overcome challenges, leverage the power of leadership, and measure the effectiveness of their cultural initiatives. Investing in creating a positive culture that aligns with the goals of finance transformation is essential for achieving long-term success in the ever-evolving world of finance.

CHAPTER 26

Finance Talent

Empowerment and Enablement

FINANCE TRANSFORMATION REFERS TO the fundamental changes taking place in the finance function of organisations to drive efficiency, effectiveness, and value creation. These transformations aim to streamline processes, leverage technology, and enhance decision-making capabilities within the finance department. The objectives of finance transformation may vary across organisations, but they typically include improving financial performance, enhancing risk management, and increasing operational agility.

Wang (2024) argues there is a shortage of technology talent in the finance industry. Wang contends professional skills in finance and emerging technologies are in a shortage with poor training rife in the industry. They argue financial institutions in particular are unable to retain talent (Wang, 2024).

As finance transformation continues to reshape the finance function, the role of finance talent becomes increasingly critical. Finance professionals are no longer limited to traditional accounting and reporting tasks; they are now expected to be strategic business partners who can provide valuable insights and support decision-making. This shift in expectations requires finance talent to possess a diverse set of skills and competencies beyond their financial acumen.

In the age of finance transformation, finance talent is no longer confined to the back office. They are actively involved in driving business growth, shaping strategies, and contributing to the overall success of the organisation. The evolving role of finance talent encompasses various dimensions, including business partnering, data analysis, technology adoption, and change management.

As business partners, finance professionals collaborate with other departments to understand their needs, provide financial insights, and help make informed decisions. They are no longer seen as mere number crunchers but as trusted advisors who can contribute to the achievement of strategic objectives. Additionally, finance talent is increasingly involved in data analysis, leveraging advanced analytics tools to extract meaningful insights from

DOI: 10.1201/9781003514503-26

vast amounts of financial data. This enables them to provide accurate forecasts, identify opportunities for cost reduction, and mitigate risks.

While the evolving role of finance talent presents exciting opportunities, it also comes with its fair share of challenges. One of the primary challenges faced by finance talent in the era of transformation is the need for continuous upskilling and reskilling. As technology continues to advance and new tools and techniques emerge, finance professionals must stay updated and acquire new skills to remain relevant.

Furthermore, the fast-paced nature of finance transformation requires finance talent to be adaptable and open to change. They must be comfortable with embracing new technologies, learning new processes, and adjusting their mindset to align with the evolving needs of the organisation. This can be challenging for individuals who are accustomed to traditional finance roles and have limited exposure to technology-driven transformations.

Another significant challenge faced by finance talent is the pressure to deliver results amidst increasing expectations. As finance professionals take on more strategic responsibilities, they are expected to provide accurate and timely insights that drive business performance. This requires them to possess strong analytical skills, attention to detail, and the ability to effectively communicate complex financial information to non-financial stakeholders.

Upskilling and reskilling are vital in developing talent. To unlock the full potential of finance talent in the age of finance transformation, organisations must invest in upskilling and reskilling initiatives. This involves providing finance professionals with the necessary training, resources, and opportunities to develop new skills and expand their knowledge base.

Upskilling focuses on enhancing existing skills and competencies, enabling finance talent to perform their current roles more effectively. This could involve training programmes on advanced financial analysis, data visualisation, or communication skills. Reskilling, on the other hand, involves acquiring entirely new skills that are relevant to the evolving needs of the finance function. This could include training on emerging technologies such as robotic process automation (RPA) or artificial intelligence.

By investing in upskilling and reskilling, organisations not only empower their finance talent to adapt to the changing landscape but also demonstrate their commitment to employee development and growth. This, in turn, helps attract and retain top finance talent who are eager to enhance their skills and contribute meaningfully to the organisation's success.

Talent attraction and retention strategies are a vital lifeline. In a highly competitive job market, attracting and retaining top finance talent can be a significant challenge for organisations. To overcome this challenge, organisations must develop strategies that not only attract finance talent but also create an environment that encourages them to stay and grow.

One strategy is to offer competitive compensation packages that align with industry standards. Finance professionals, especially those with specialised skills, are in high demand, and organisations must offer attractive compensation to remain competitive. Additionally, organisations should focus on creating a positive work culture that fosters collaboration, innovation, and continuous learning. This includes providing opportunities for career progression, recognising and rewarding exceptional performance, and promoting work-life balance.

Moreover, organisations should leverage their employer brand to attract top finance talent. This involves showcasing the organisation's commitment to finance transformation, highlighting the exciting opportunities available, and emphasising the organisation's values and culture. By effectively communicating the organisation's vision and mission, organisations can attract finance talent who are aligned with their goals and aspirations.

Technology plays a crucial role in enhancing the productivity of finance talent in the age of finance transformation. By leveraging automation, artificial intelligence, and advanced analytics tools, finance professionals can streamline repetitive tasks, reduce manual errors, and focus on more value-added activities.

For example, the implementation of RPA can automate routine finance processes such as invoice processing or reconciliations. This not only improves efficiency but also frees up finance talent to focus on strategic analysis and decision-making. Similarly, advanced analytics tools enable finance professionals to extract meaningful insights from large volumes of financial data, enabling them to identify trends, risks, and opportunities.

Additionally, cloud-based technologies have revolutionised the way finance professionals collaborate and access information. Cloud-based finance systems allow real-time data sharing, enabling finance talent to make informed decisions quickly. This promotes agility and responsiveness, essential qualities in a rapidly changing business environment.

Data and analytics have become indispensable tools for optimising finance talent performance. By leveraging data, finance professionals can gain valuable insights into their own performance and identify areas for improvement. This includes analysing key performance indicators (KPIs) such as the accuracy and timeliness of financial reporting, the efficiency of financial processes, and the effectiveness of decision-making.

Furthermore, data and analytics can be used to identify patterns and trends in finance talent performance across the organisation. This allows organisations to identify top performers, provide targeted coaching and development opportunities, and drive continuous improvement. By leveraging data, organisations can make data-driven decisions regarding talent allocation, training needs, and performance management.

Developing a *culture of continuous learning and improvement* is essential for unlocking the potential of finance talent in the age of finance transformation. Organisations can foster this culture by implementing the following best practices:

1. Encourage a Growth Mindset: Promote the belief that talent can be developed through effort and dedication. Encourage finance professionals to embrace challenges, learn from failures, and continuously seek opportunities for growth.

2. Provide Learning Opportunities: Offer a variety of learning opportunities, including training programmes, webinars, workshops, and conferences. Encourage finance talent to pursue certifications or advanced degrees to enhance their knowledge and skills.

3. Foster Knowledge Sharing: Create platforms for finance professionals to share their expertise and learn from their peers. This can include knowledge-sharing sessions, cross-functional projects, or mentoring programmes.

4. Embrace Technology-enabled learning: Leverage technology to provide flexible and accessible learning opportunities. This could include e-learning platforms, mobile apps, or virtual classrooms.

5. Recognise and Reward Learning: Acknowledge and reward finance professionals who actively engage in learning and demonstrate a commitment to continuous improvement. This can include promotions, bonuses, or special assignments.

By implementing these best practices, organisations can create a culture that values learning, encourages innovation, and supports the development of finance talent.

Case studies help us consider how we can better leverage finance talent. These case studies illustrate how organisations can leverage finance talent in the age of finance transformation to drive positive outcomes and create a competitive advantage:

1. *Microsoft*: Microsoft, a multinational technology company, recognised the importance of finance transformation and invested in upskilling its finance talent. They provided training on advanced data analytics tools and fostered a culture of data-driven decision-making. As a result, their finance talent was able to provide valuable insights that led to cost savings, improved forecasting accuracy, and enhanced risk management.

2. *Toyota Group*: Toyota Group, a global manufacturing company, implemented RPA to streamline their finance processes. This allowed their finance talent to focus on more strategic activities such as business partnering and analysis. The implementation of RPA not only improved efficiency but also increased job satisfaction among finance professionals.

Finance transformation is reshaping the finance function, and finance talent plays a crucial role in driving the success of these transformations. To unlock the full potential of finance talent, organisations must understand the evolving role of finance professionals, address the challenges they face, and invest in upskilling and reskilling initiatives.

By attracting and retaining top finance talent, leveraging technology, optimising performance through data and analytics, and fostering a culture of continuous learning, organisations can maximise the potential of their finance talent in the age of finance transformation. The successful examples highlighted in the case studies demonstrate that organisations that invest in their finance talent reap the rewards of improved financial performance, increased efficiency, and better decision-making.

Organisations that prioritise the development and empowerment of their finance talent are well-positioned to thrive in the age of finance transformation. By recognising the importance of finance talent and providing the necessary support and resources, organisations can navigate the challenges of transformation and drive sustainable growth and success.

CHAPTER 27

Motivating Your Finance Team Through Transformation

Finance transformation is a complex process that requires the collaboration and dedication of every member of your team. However, without proper motivation, your team may struggle to embrace the changes and adapt to new ways of working. Motivation plays a crucial role in driving your finance team towards successful transformation. It ensures that they remain focused, enthusiastic, and committed to achieving the organisation's goals. Chief Finance Officer (CFO) must consider the importance of motivation in finance transformation and how it can impact the overall success of their team.

Motivation is the driving force that propels individuals and teams to overcome obstacles, embrace change, and strive for excellence. Motivated teams are more likely to embrace change and willingly participate in the finance transformation process. When employees are motivated, they are more engaged, proactive, and willing to go the extra mile to achieve their objectives. They are driven by a sense of purpose and understand the value they bring to the organisation. Motivated team members are also more resilient in the face of challenges and setbacks, as they believe in their ability to overcome obstacles and make a positive impact.

To motivate your finance team, it is essential to clearly communicate the purpose and benefits of the finance transformation. Help them understand how their efforts contribute to the organisation's success and how their roles will evolve as part of the transformation. By providing a clear vision and explaining the 'why' behind the changes, you can inspire your team and increase their motivation level.

Before delving into strategies for motivating finance teams, it is important to understand the unique challenges and motivations that they face. Finance professionals often encounter demanding workloads, stringent deadlines, and high levels of pressure. They are responsible for managing financial risks, ensuring compliance, and providing accurate and timely financial information. These challenges can sometimes lead to stress, burnout, and a lack of motivation. However, finance teams are also driven by a desire

DOI: 10.1201/9781003514503-27

for professional growth, recognition, and the opportunity to make a meaningful impact on the organisation. By recognising and addressing these challenges and motivations, leaders can create an environment that fosters motivation and empowers finance teams to achieve transformational goals.

Finance transformation brings about significant changes in processes, systems, and roles within the finance function. While these changes are necessary for growth and improvement, they can also create challenges and resistance within your team. It is important to understand these challenges and address them effectively to maintain motivation and ensure a smooth transformation process.

One of the common challenges in finance transformation is the fear of job displacement. Employees may feel anxious about the introduction of new technologies and automation, fearing that it will make their roles redundant. It is crucial to address these concerns and provide reassurance that the transformation is not about replacing jobs but rather about enhancing efficiency and adding value to the organisation. Emphasise the importance of upskilling and reskilling, highlighting the opportunities for growth and career advancement.

Another challenge is the resistance to change. Change can be uncomfortable and unsettling for many individuals, leading to a lack of motivation and engagement. To overcome this challenge, involve your finance team in the decision-making process and encourage their input. By giving them a sense of ownership and control, they will be more likely to embrace the changes and actively participate in the transformation. Provide training and support to help them adapt to new processes and technologies, and address any concerns or questions they may have along the way.

Effective leadership plays a crucial role in motivating finance teams and driving success during finance transformation. Leaders have the ability to inspire, influence, and empower their teams through their actions and behaviours. One key aspect of effective leadership is the ability to communicate a compelling vision and purpose. By clearly articulating the goals and objectives of the finance transformation, leaders can create a sense of purpose and inspire their teams to work towards a common goal. In addition, leaders should provide ongoing feedback and support to their team members. Regular communication and feedback sessions can help identify any challenges or roadblocks and provide the necessary guidance and support to overcome them. By demonstrating trust, respect, and appreciation for their team members, leaders can create a positive and motivating work environment that fosters collaboration and innovation.

Motivation is not a one-size-fits-all approach. Different individuals are motivated by different factors. As a leader, it is essential to understand the unique needs and preferences of your finance team members to effectively motivate and inspire them. CFOs must consider different strategies that can help them create a motivated and high-performing finance team during the transformation process.

Setting clear and measurable goals is a fundamental step in motivating finance teams to achieve transformational goals. When goals are vague or ambiguous, it can be difficult for team members to understand what is expected of them and how their efforts contribute to the overall success of the transformation. Clear goals provide clarity and direction, allowing individuals to align their efforts and prioritise their work accordingly. In addition, goals should be measurable, meaning that progress can be tracked and milestones can be celebrated. This not only provides a sense of accomplishment but also helps to maintain motivation and momentum throughout the transformation process. By involving the finance team in the goal-setting process and ensuring that goals are challenging yet attainable, leaders can foster a sense of ownership and commitment among team members.

Clear goals and expectations provide your finance team with a sense of direction and purpose. When team members have a clear understanding of what is expected of them and the goals they are working towards, they are more likely to stay motivated and focused. Ensure that the goals are SMART (specific, measurable, achievable, relevant, and time-bound) and communicate them effectively to your team. Regularly review progress and provide feedback to keep them on track and motivated.

Investing in the *training and development* of your finance team is crucial for their motivation and growth. Offer *continuous learning opportunities* to enhance their skills and knowledge in areas relevant to the finance transformation. This can include workshops, seminars, online courses, or mentorship programs. Encourage your team members to take ownership of their professional development and provide them with the necessary resources and support to succeed.

Inclusion is an essential component of motivating finance teams for transformational goals. An inclusive work environment is one that embraces diversity, values different perspectives, and ensures equal opportunities for all team members. When individuals feel included and appreciated for their unique contributions, they are more likely to be motivated and engaged. Leaders can promote inclusivity by creating a culture of respect, empathy, and fairness. This can be achieved through inclusive hiring practices, diversity training, and the establishment of policies and procedures that promote equal opportunities. By fostering an inclusive work environment, leaders can tap into the full potential of their finance teams and drive success during finance transformation.

A *positive work environment* is essential for maintaining motivation and productivity. Foster a culture of trust, respect, and open communication within your finance team. Encourage collaboration and teamwork, celebrate achievements, and provide constructive feedback.

Recognise and appreciate the efforts of your team members regularly. Create a supportive atmosphere where individuals feel valued, motivated, and comfortable expressing their ideas and concerns.

Collaboration and recognition are powerful motivators that can drive success during finance transformation. When team members feel valued, included, and respected, they are more likely to be motivated and engaged. Leaders can foster a culture of collaboration by encouraging open communication, teamwork, and knowledge sharing. This can be achieved through regular team meetings, cross-functional projects, and the use of collaborative tools and technologies. In addition, leaders should recognise and celebrate the achievements and contributions of their team members. Whether through verbal praise, public recognition, or tangible rewards, acknowledging the hard work and dedication of individuals can boost morale, increase motivation, and reinforce a positive work culture.

Incentives and rewards can be powerful tools for motivating finance teams to achieve transformational goals. While financial incentives such as bonuses and salary increases can be effective, leaders should also consider non-financial rewards that align with the individual preferences and motivations of their team members. This can include opportunities for career development, flexible work arrangements, or recognition programs. By tailoring incentives and rewards to the needs and aspirations of their team members, leaders can create a motivating and fulfilling work environment that encourages high performance and continuous improvement.

Acknowledging and rewarding achievements is a powerful way to motivate and inspire your finance team. Celebrate milestones, both big and small, and publicly recognise the contributions of individuals and the team as a whole. Rewards can be in the form of monetary incentives, promotions, public recognition, or even simple gestures like a handwritten note of appreciation. By recognising their efforts, you reinforce a culture of excellence and motivate your team to strive for continued success.

Effective communication is vital throughout the finance transformation process. It ensures that your team members are well-informed, engaged, and aligned with the objectives of the transformation. In this section, we will explore the importance of effective communication and strategies to facilitate clear and open communication within your finance team.

Clear and transparent communication is essential to mitigate uncertainty and manage expectations during finance transformation. Regularly communicate updates, progress, and changes to your team members. Use a variety of communication channels, such as team meetings, emails, newsletters, and intranet portals, to ensure that information reaches everyone in a timely manner. Encourage open dialogue and provide opportunities for your team members to ask questions, seek clarification, and share their thoughts and concerns.

It is also important to tailor your communication to the needs and preferences of your team members. Some individuals may prefer face-to-face interactions, while others may prefer written communication. Take the time to understand their communication styles and adapt your approach accordingly. Be empathetic and approachable, creating a safe space for your team members to express their opinions and concerns without fear of judgement.

A *strong team culture* is a powerful motivator for your finance team during the transformation process. When team members feel connected, supported, and valued, they are more likely to be motivated and engaged. In this section, we will explore strategies to build a strong team culture within your finance team.

Encourage collaboration and teamwork by fostering an environment where ideas are openly shared and respected. Create cross-functional teams to tackle specific challenges and promote knowledge sharing across different areas of expertise. Provide opportunities for team members to work on joint projects and initiatives, allowing them to learn from each other and develop a sense of camaraderie.

Promote a culture of continuous improvement by encouraging feedback and embracing a growth mindset. Create channels for anonymous feedback to ensure that everyone feels comfortable voicing their opinions and suggestions. Regularly review and implement the feedback received, demonstrating that their input is valued and contributes to positive change.

To illustrate effective strategies for motivating finance teams, let's explore a few case studies of organisations that have successfully achieved their transformational goals through motivation and empowerment.

Boston Consulting Group: Boston Consulting Group embarked on a finance transformation journey to streamline their financial processes and improve efficiency. The leaders of the finance team recognised the importance of motivation and empowerment and implemented a series of initiatives to foster a culture of collaboration, recognition, and inclusivity. By setting clear and measurable goals, providing ongoing feedback and support, and offering incentives and rewards, the finance team was able to successfully achieve their transformational goals and drive positive change within the organisation.

Salesforce: Salesforce faced a significant challenge in implementing a new financial system as part of their finance transformation. The leaders of the finance team recognised the potential resistance and lack of motivation among team members due to the complexity and disruption caused by the system implementation. To overcome this challenge, they implemented a comprehensive change management programme that included clear communication, training, and support. By addressing the concerns and anxieties of team members and providing the necessary resources and guidance, the finance team was able to overcome the initial resistance and successfully implement the new financial system.

Measuring and evaluating success with team motivation during finance transformation is essential to ensure continuous improvement and drive long-term success. Leaders should establish key performance indicators (KPIs) that align with the goals and objectives of the finance transformation. These KPIs can include metrics such as employee engagement, productivity, and the achievement of transformational goals. Regular monitoring and evaluation of these KPIs can provide insights into the effectiveness of the motivation strategies and help identify areas for improvement. In addition, leaders should seek feedback from their team members through surveys, focus groups, or one-on-one discussions. By actively listening to the concerns, suggestions, and feedback of their team members, leaders can make informed decisions and adjustments to their motivation strategies, ensuring continuous improvement and sustained success.

Motivating finance teams to achieve transformational goals requires a combination of effective leadership, clear goals, collaboration, recognition, inclusivity, and the use of incentives and rewards. By understanding the unique challenges and motivations of finance teams, leaders can create an empowering work environment that fosters motivation, engagement, and high performance. Through the implementation of effective motivation strategies and continuous evaluation, leaders can drive success during finance transformation and propel their organisations towards a brighter future.

Motivating and inspiring your finance team is a critical aspect of successful finance transformation. By understanding the importance of motivation, addressing challenges, and implementing effective strategies, you can create a motivated and high-performing team that drives the transformation process forward. Remember to set clear goals, provide ongoing training, create a positive work environment, recognise achievements, communicate effectively, and build a strong team culture. By doing so, you will unleash the true potential of your finance team and achieve a successful transformation.

CHAPTER 28

How Diversity in Finance Talent Can Drive Transformation

IN TODAY'S RAPIDLY CHANGING business landscape, diversity has become a critical factor in the success of any organisation, particularly in the finance industry. Embracing diversity goes beyond meeting quotas and creating a politically correct image – it is about tapping into the full potential of a diverse workforce. A diverse team brings together different perspectives, experiences, and skills, leading to more innovative ideas and better decision-making.

Robert (2024) argues the strategic importance of diversity and inclusion (D&I) in talent acquisition strategies, and these have even more impact in financial services firms or finance functions due to the finance transformation talent shortage (Robert, 2024).

Financial institutions that prioritise diversity are more likely to attract top talent and gain a competitive advantage. Studies have shown that diverse teams outperform homogeneous teams, resulting in higher profitability and better long-term business outcomes. Furthermore, a diverse workforce can better understand and serve a wide range of clients, leading to increased customer satisfaction and loyalty.

Cultivating a diverse and inclusive team in the finance industry offers numerous benefits. Firstly, diverse teams bring a variety of perspectives, enabling them to analyse problems from different angles and develop creative solutions. This can lead to more effective risk management and improved decision-making processes, ultimately enhancing the overall financial performance of the organisation.

Secondly, a diverse and inclusive team in finance fosters a culture of innovation. When individuals from various backgrounds come together, they bring unique ideas and insights that can drive the development of new products, services, and strategies. This innovation can help financial institutions stay ahead of the competition and adapt to the evolving needs of their clients and the market.

Moreover, a diverse team in finance can better understand and serve a diverse customer base. Different cultures, ethnicities, and genders have unique financial needs and

DOI: 10.1201/9781003514503-28

preferences. By having a team that reflects the diversity of their customers, financial institutions can build trust and establish stronger relationships, leading to increased customer satisfaction and loyalty.

Despite the many benefits of diversity in finance, there are several challenges that organisations face in achieving it. One of the main challenges is the lack of representation and opportunities for marginalised groups. Historically, the finance industry has been predominantly male dominated, with limited representation of women, racial minorities, and individuals from disadvantaged backgrounds.

Another challenge is unconscious bias in the hiring and promotion processes. Unconscious biases can lead to the preference of candidates who are similar to those already in power, perpetuating the lack of diversity within organisations. Overcoming unconscious bias requires implementing objective and inclusive hiring and promotion practices, such as blind resume screening and diverse interview panels.

Additionally, the finance industry has traditionally valued certain educational backgrounds and experiences, which can create barriers for individuals from non-traditional backgrounds.

To overcome this challenge, organisations need to broaden their definition of talent and actively seek out individuals with diverse skills, experiences, and perspectives.

To build a diverse and inclusive team in finance, organisations need to adopt *effective strategies* for hiring and promoting diverse talent. One strategy is to implement targeted recruitment initiatives that focus on attracting candidates from underrepresented groups. This can include partnering with organisations that support diverse talent, attending career fairs specifically targeted at diverse candidates, and offering internships and mentorship programmes.

Another strategy is to ensure that the hiring and promotion processes are fair and unbiased. This can be achieved by implementing blind resume screening, where personal information such as names, gender, and ethnicity are removed from resumes before they are reviewed. Additionally, organisations should strive to have diverse interview panels to eliminate unconscious bias and ensure that candidates from different backgrounds are given equal opportunities.

Organisations can also create development and mentorship programmes to support the growth and advancement of diverse talent. These programmes can provide training, networking opportunities, and access to senior leaders who can provide guidance and support. By investing in the development of diverse talent, organisations can create a pipeline of future leaders who will contribute to the long-term success of the organisation.

Creating a truly *inclusive* workplace is essential for unleashing the full power of diversity in finance. Inclusion goes beyond simply having diverse individuals within the organisation; it is about creating an environment where everyone feels valued, respected, and empowered to contribute their unique perspectives and ideas.

One way to foster inclusivity is by promoting open and transparent communication. Encouraging employees to share their thoughts and opinions freely creates an environment where diverse perspectives are welcomed and valued. Additionally, organisations should provide training and education on D&I to raise awareness and promote understanding among employees.

Another important aspect of fostering inclusivity is ensuring equal opportunities for growth and development. This can be achieved by offering mentorship programmes, providing training and development opportunities, and creating a culture of feedback and recognition. By investing in the professional growth of all employees, organisations can create a sense of belonging and empower individuals to reach their full potential.

Creating a *culture of D&I* requires a long-term commitment from the organisation's leadership. Leaders play a crucial role in setting the tone and creating an environment where D&I are valued and prioritised. They must lead by example, demonstrating inclusive behaviours and actively promoting diversity in all aspects of the organisation's operations.

Leaders can also establish D&I metrics and hold themselves and others accountable for achieving them. By tracking progress and regularly reporting on D&I initiatives, leaders can ensure that the organisation remains committed to its goals and continuously works towards creating a more diverse and inclusive workplace.

Furthermore, leaders should actively seek out diverse perspectives and ideas in decision-making processes. This can be achieved by forming diverse teams and seeking input from individuals with different backgrounds and experiences. By incorporating diverse perspectives into decision-making, leaders can make more informed and inclusive choices that benefit the organisation as a whole.

Leadership plays a critical role in driving D&I in the finance industry. As the driving force behind organisational culture and values, leaders have the power to shape the way diversity is perceived and embraced within the organisation. Without strong leadership support, D&I initiatives are likely to stall and be seen as mere lip service.

Leaders can demonstrate their commitment to D&I by setting clear goals and expectations for the organisation. They should communicate the importance of D&I to all employees and ensure that it is integrated into the organisation's mission, vision, and values. By aligning D&I with the organisation's strategic objectives, leaders can create a sense of purpose and urgency around these initiatives.

Furthermore, leaders should invest in their own development and education on D&I. This can include attending workshops and conferences, reading relevant literature, and seeking out diverse perspectives. By educating themselves, leaders can gain a deeper understanding of the challenges and opportunities associated with D&I and become effective advocates for change.

Several financial institutions have successfully implemented diversity initiatives and created inclusive workplaces. One such example is *American Express*, which has been recognised for its commitment to D&I. The company has established diversity councils, employee resource groups, and mentoring programmes to support the growth and development of diverse talent. As a result, American Express has seen increased employee engagement and satisfaction and has been able to attract top talent from diverse backgrounds.

Another example is *JPMorgan Chase*, which has implemented a comprehensive D&I strategy. The company has set specific goals for increasing the representation of women and minorities in senior leadership positions and has implemented unconscious bias training for all employees. JPMorgan Chase has also established partnerships with organisations that support diverse talent and has created programmes to provide access to capital for minority-owned businesses.

Tools and resources are a vital lifeline. Implementing diversity initiatives and driving finance transformation requires access to the right tools and resources. Fortunately, there are several organisations and platforms that provide support and guidance in these areas. One such resource is Catalyst, a global non-profit that works with organisations to build inclusive workplaces and advance women in leadership. Catalyst offers research, tools, and programmes to help organisations drive D&I.

Another valuable resource is the Diversity and Inclusion Consortium for the Finance Industry (DICFI). DICFI is a collaborative platform that brings together financial institutions, industry experts, and thought leaders to share best practices and drive D&I in the finance industry. The consortium provides access to a network of professionals, educational resources, and events focused on D&I.

Additionally, organisations can leverage technology to support their D&I initiatives. There are several software platforms available that can help track diversity metrics, analyse workforce data, and identify areas for improvement. These tools can provide valuable insights and help organisations measure the effectiveness of their diversity initiatives.

Diversity plays a crucial role in finance transformation. As the finance industry continues to evolve, organisations need to adapt to changing customer expectations, technological advancements, and regulatory requirements. A diverse and inclusive team is better equipped to navigate these challenges and drive innovation and growth.

By embracing diversity, financial institutions can attract and retain top talent, enhance decision-making processes, and better understand and serve their customers. This can lead to improved financial performance, increased customer satisfaction, and a stronger competitive advantage.

Unleashing the power of diversity in finance is not only the right thing to do, but it is also a strategic imperative. By cultivating a talented and inclusive team, financial institutions can drive innovation, improve financial performance, and build strong relationships with their customers. However, achieving diversity in finance requires a long-term commitment, strong leadership, and the implementation of effective strategies. By embracing diversity and creating a culture of inclusion, the finance industry can truly unlock its full potential and thrive in the modern business landscape.

CHAPTER 29

Research with Female Finance Leaders and Empowering Talent

T HE WORLD OF FINANCE has traditionally been dominated by men, but in recent years, we have seen a shift towards greater gender diversity in leadership roles. Female finance leaders are playing a crucial role in driving finance transformation within organisations. They bring a unique perspective and a fresh approach to solving complex financial challenges. With their strong analytical skills, strategic thinking, and ability to adapt to changing market conditions, female finance leaders are breaking barriers and leading the way in finance transformation.

Girardone et al. (2021) argue the critical nature of the importance of ethnic diversity and female representation at senior management and board level in finance. They argue diversity, including gender diversity, is vital to effective decision-making, talent retention, and financial performance (Girardone et al., 2021).

One of the key roles of female finance leaders in finance transformation is fostering a culture of innovation and change. They understand the importance of embracing new technologies and processes to stay competitive in today's fast-paced business environment. By championing digital transformation initiatives, they are able to streamline financial operations, improve efficiency, and enhance decision-making processes. Female finance leaders also prioritise collaboration and teamwork, creating a supportive and inclusive environment where everyone's ideas are valued. This collaborative approach is essential in driving finance transformation and ensuring its success.

While female finance leaders are making significant strides in finance transformation, they still face unique challenges along the way. One of the biggest challenges is breaking through the glass ceiling and gaining access to top leadership positions. Despite their qualifications and expertise, women often face bias and discrimination when it comes to career advancement. This can result in a lack of representation at the highest levels of

DOI: 10.1201/9781003514503-29

finance leadership, limiting the opportunities for female finance leaders to drive meaningful change.

Another challenge faced by female finance leaders is the work-life balance dilemma. Balancing demanding careers with family responsibilities can be particularly challenging for women. This can sometimes hinder their career progression and limit their ability to fully dedicate themselves to driving finance transformation. Organisations need to prioritise creating a supportive and flexible work environment that allows female finance leaders to thrive both personally and professionally.

Success stories of female finance leaders are vital in raising aspirations and breaking down stereotypes. Despite the challenges they face, many female finance leaders have achieved remarkable success in driving finance transformation within their organisations. One such success story is that of *Jane Thompson*, the Chief Finance Officer (CFO) of a multinational corporation. Under her leadership, the company underwent a comprehensive finance transformation initiative that resulted in significant cost savings and improved financial performance. Jane's strategic vision, strong leadership skills, and ability to rally her team were instrumental in the success of this transformation.

Another inspiring success story is *Sarah Patel*, the finance director of a tech startup. Sarah played a key role in transforming the company's financial operations by implementing advanced analytics and automation tools. Her innovative approach not only improved efficiency but also provided valuable insights that enabled the company to make data-driven decisions. Sarah's ability to embrace technology and leverage data has positioned her as a trailblazer in finance transformation.

These success stories highlight the immense potential of female finance leaders in driving finance transformation. Their unique perspectives, collaborative leadership styles, and innovative thinking are invaluable assets in today's rapidly changing business landscape.

Diversity in finance talent is a critical factor in driving finance transformation. When organisations have a diverse range of backgrounds, experiences, and perspectives within their finance teams, they are better equipped to tackle complex challenges and identify innovative solutions. Female finance leaders bring a different set of skills and perspectives to the table, which can lead to more robust and well-rounded financial strategies.

Studies have shown that companies with gender-diverse leadership teams outperform their peers in terms of financial performance. This is because diverse teams are more likely to consider a wider range of possibilities, challenge the status quo, and make better decisions. By promoting diversity in finance talent, organisations can tap into a wealth of untapped potential and drive finance transformation to new heights.

Strategies for promoting gender diversity play an important role. To promote gender diversity in finance leadership roles, organisations need to adopt proactive strategies that address the barriers faced by female finance professionals. One such strategy is implementing mentoring and sponsorship programs. These programmes pair female finance professionals with senior leaders who can provide guidance, support, and advocacy. Mentoring and sponsorship programmes not only help women navigate their careers but also create a pipeline of future female finance leaders.

Another effective strategy is implementing flexible work arrangements. By offering flexible schedules, remote work options, and parental leave policies, organisations can support the work-life balance of female finance professionals. This not only improves retention rates but also attracts top talent to finance leadership roles.

Organisations can also promote gender diversity by providing equal opportunities for training and development. By investing in the professional growth of female finance professionals, organisations can nurture their skills and talents, preparing them for future leadership positions.

Key skills and qualities must also be considered. Successful female finance leaders possess a unique set of skills and qualities that enable them to drive finance transformation. One key skill is strong analytical ability. Female finance leaders are adept at analysing complex financial data, identifying trends, and making informed decisions. This analytical prowess allows them to navigate the ever-changing financial landscape and drive strategic transformation initiatives.

Another important quality of successful female finance leaders is effective communication. They are skilled at translating complex financial concepts into clear and concise messages that resonate with stakeholders at all levels of the organisation. This ability to communicate effectively ensures that finance transformation initiatives are understood and embraced by all.

Finally, resilience is a critical quality for female finance leaders. They face numerous challenges and obstacles in their journey towards finance transformation. Resilience allows them to overcome setbacks, learn from failures, and continue driving change even in the face of adversity.

Resources and organisations can help. There are numerous resources and organisations dedicated to supporting and empowering female finance leaders. One such resource is the Women in Finance Network, which provides a platform for networking, mentoring, and professional development. This network offers access to a vibrant community of female finance professionals who can provide guidance and support.

Organisations such as Lean In and Catalyst also offer valuable resources and programmes focused on promoting gender diversity in leadership roles. These organisations provide research, tools, and best practices for organisations looking to create more inclusive and diverse finance teams.

It is encouraging to see the growing number of resources and organisations dedicated to supporting female finance leaders. By leveraging these resources, organisations can create a more inclusive and equitable finance landscape.

Several companies have experienced successful finance transformation under the leadership of female finance leaders. One such case is *Zoom*, where *Kelly Steckelberg*, the CFO, spearheaded a comprehensive finance transformation initiative. By implementing cutting-edge technology, streamlining processes, and enhancing financial reporting, Kelly was able to drive significant cost savings and improve the overall financial performance of the company.

Another inspiring case study is *Barclays Bank*, where *Anna Cross*, the Group Finance Director, led a finance transformation initiative focused on enhancing risk management and regulatory compliance. Through her strategic vision and collaborative leadership,

Anna successfully implemented new processes that improved the bank's risk profile and ensured compliance with industry regulations.

These case studies demonstrate the impact that female finance leaders can have on finance transformation. Their ability to drive change, foster innovation, and create a culture of continuous improvement can lead to substantial benefits for organisations.

As finance transformation continues to gain momentum, the role of female finance leaders will become increasingly crucial. Their unique perspectives, collaborative leadership styles, and innovative thinking are essential in driving meaningful change and ensuring the long-term success of finance transformation initiatives.

To fully leverage the potential of female finance leaders, organisations must address the barriers they face and create a supportive and inclusive environment. By promoting gender diversity, offering equal opportunities for advancement, and providing resources and support, organisations can tap into a diverse pool of talent and drive finance transformation to new heights.

The future of finance transformation is bright, and female finance leaders are at the forefront of this exciting journey. Let us embrace their contributions, break down barriers, and empower more women to take on leadership roles in finance transformation. Together, we can create a more inclusive and dynamic financial landscape for the benefit of organisations and society as a whole.

CHAPTER 30

How Aligning Digital Strategies Can Drive Business Outcomes

In today's rapidly evolving business landscape, digital strategies have become essential for organisations across industries, including finance. Embracing digital transformation is no longer an option but a necessity for financial institutions to stay competitive and maximise their financial success. Digital strategies encompass a range of technologies, tools, and approaches that enable businesses to streamline operations, enhance decision-making, and drive better business outcomes.

Digital strategies in finance enable financial institutions to leverage technology to optimise their operations, improve customer experiences, and drive growth. By embracing digital transformation, organisations can gain a competitive advantage, expand their reach, and provide innovative solutions to their customers.

Huang et al. (2024) argue businesses must choose digital strategies that suit their particular characteristics. These strategies encompass various areas, including data analytics, automation, artificial intelligence (AI), and digital marketing, among others. By aligning these strategies with business outcomes, financial institutions can unlock new opportunities and drive their success to new heights (Huang and Gao, 2024).

Aligning digital strategies with business outcomes has numerous benefits for financial institutions. Firstly, it helps in streamlining operations and improving efficiency. By leveraging automation and AI, financial institutions can automate repetitive tasks, reduce manual errors, and improve operational efficiency. This not only saves time and resources but also allows employees to focus on more strategic tasks, driving better business outcomes.

Secondly, aligning digital strategies with business outcomes enables better decision-making through data analytics. By leveraging data analytics tools and techniques, financial institutions can gain valuable insights into customer behaviour, market trends, and risk

DOI: 10.1201/9781003514503-30

assessment. These insights empower organisations to make informed decisions, develop tailored products and services, and identify new growth opportunities.

Moreover, aligning digital strategies with business outcomes enhances customer experiences. Through digital marketing techniques, financial institutions can reach their target audience more effectively, personalise their offerings, and provide seamless digital experiences. This not only improves customer satisfaction but also builds long-term customer relationships, leading to higher retention rates and increased business outcomes.

Digital strategies are vital. To maximise financial success, financial institutions need to adopt key digital strategies. These strategies encompass various areas and technologies, each contributing to different aspects of the business. Here are some key digital strategies that can drive business outcomes in finance:

Leveraging Data Analytics for Better Financial Decision-Making. Data analytics is a powerful tool for financial institutions to make better financial decisions. By analysing large volumes of data, organisations can identify patterns, trends, and insights that can inform investment decisions, risk assessments, and customer segmentation. Advanced analytics techniques, such as predictive modelling and machine learning, can help financial institutions gain a competitive edge by identifying emerging opportunities and mitigating potential risks.

Implementing Automation and AI in Finance for Improved Efficiency. Automation and AI technologies have the potential to revolutionise the finance industry by improving efficiency and reducing costs. By automating manual tasks, such as data entry and reconciliation, financial institutions can free up resources and enable employees to focus on more value-added activities. AI-powered algorithms can also enhance fraud detection, credit scoring, and investment recommendations, leading to improved business outcomes and customer satisfaction.

Digital marketing plays a crucial role in driving business outcomes for financial institutions. Through digital channels, organisations can reach a wider audience, engage with customers in real-time, and create personalised experiences. By leveraging digital marketing techniques, such as search engine optimisation (SEO), social media marketing, and content marketing, financial institutions can increase brand awareness, generate leads, and drive customer acquisition.

Effectively aligning digital strategies is important. While aligning digital strategies with business outcomes can be a game-changer for financial institutions, it requires careful planning and execution. Here are some tips to effectively align digital strategies with business outcomes in finance:

1. Set Clear Business Goals: Define specific business outcomes that you want to achieve through digital strategies, such as increasing revenue, improving customer satisfaction, or reducing costs. These goals will serve as a guiding framework for your digital transformation initiatives.

2. Invest in the Right Technologies: Identify the digital technologies and tools that align with your business goals. Invest in robust data analytics platforms, automation software, and AI solutions that can address your organisation's unique needs and drive desired outcomes.

3. Foster a Culture of Innovation: Encourage a culture of innovation and continuous improvement within your organisation. Empower employees to explore new ideas, embrace emerging technologies, and experiment with digital strategies to drive business outcomes.

4. Prioritise Data Security and Privacy: As financial institutions deal with sensitive customer information, it is crucial to prioritise data security and privacy. Implement robust security measures, adhere to data protection regulations, and invest in secure infrastructure to protect confidential data.

Implementing digital strategies in finance requires the right *tools and resources*. Here are some essential tools and resources that can help financial institutions drive their digital transformation:

- Data Analytics Platforms: Tools like Tableau, Power BI, and Google Analytics enable financial institutions to analyse data, gain insights, and make informed business decisions.

- Automation Software: Platforms like UiPath, Automation Anywhere, and Blue Prism automate manual tasks, streamline workflows, and improve operational efficiency.

- AI Solutions: Technologies like IBM Watson, Amazon SageMaker, and Google Cloud AI provide AI-powered capabilities, such as natural language processing, predictive analytics, and image recognition.

- Digital Marketing Platforms: Tools like HubSpot, Google Ads, and Hootsuite facilitate digital marketing activities, such as social media management, email marketing, and content creation.

Continued alignment plays a crucial role. As technology continues to advance, the future of finance will be increasingly digital. Financial institutions must continue to align their strategies with digital transformation to stay relevant and thrive in the evolving landscape. Embracing emerging technologies, such as blockchain, Internet of Things (IoT), and quantum computing, will be crucial for financial institutions to drive innovation, improve customer experiences, and achieve long-term financial success.

Aligning digital strategies with business outcomes is imperative for financial institutions to maximise their financial success. By leveraging data analytics, automation, AI, and digital marketing, organisations can streamline operations, enhance decision-making, and drive better business outcomes. The future of finance lies in embracing digital transformation and continuously aligning strategies with emerging technologies. By doing so, financial institutions can stay competitive, provide superior customer experiences, and unlock new opportunities for growth.

As the financial landscape continues to evolve, the role of *Chief Financial Officers (CFOs)* has become increasingly crucial in driving business growth. No longer confined to the traditional responsibilities of financial planning and reporting, CFOs are now expected to play a strategic role in shaping the future of their organisations. By leveraging digital strategies, CFOs can unlock new opportunities for growth and position their companies for success in the digital age.

CFOs have always been responsible for managing the financial health of their organisations. This includes overseeing budgeting, forecasting, and financial reporting. However, the modern CFO is now expected to go beyond these traditional functions and actively contribute to the strategic direction of the company. By understanding the business landscape and aligning financial strategies with broader business objectives, CFOs can drive growth and create value for their organisations.

Digital strategies have had a profound impact on the field of finance. The emergence of new technologies has enabled CFOs to streamline financial processes, improve efficiency, and make more informed decisions. By leveraging automation, artificial intelligence, and data analytics, CFOs can gain deeper insights into their organisation's financial performance and identify areas for improvement.

One of the key benefits of digital strategies in finance is the ability to gather and analyse vast amounts of data. With the right tools and technologies in place, CFOs can access real-time financial information and make data-driven decisions. This not only improves the accuracy of financial reporting but also enables CFOs to identify trends and patterns that can inform strategic decision-making.

Aligning digital strategies is key. To effectively drive business growth through digital strategies, CFOs must ensure that these strategies are aligned with broader business outcomes. It is not enough to simply adopt new technologies; CFOs must understand how these technologies can contribute to the organisation's goals and objectives.

By aligning digital strategies with business outcomes, CFOs can prioritise investments and allocate resources effectively. For example, if the goal is to increase operational efficiency, CFOs may choose to invest in automation technologies that streamline financial processes. If the goal is to improve customer experience, CFOs may invest in technologies that enable personalised financial services.

Digital strategies offer CFOs the opportunity to create *new products and services* that can drive business growth. By leveraging technology and data, CFOs can identify customer needs and develop innovative solutions to meet those needs.

For example, digital payment platforms have revolutionised the way we conduct financial transactions. CFOs can leverage these platforms to create new revenue streams and provide customers with convenient and secure payment options. By embracing digital strategies, CFOs can stay ahead of the competition and position their organisations as industry leaders.

Digital transformation presents CFOs with a myriad of *opportunities for growth*. By embracing digital strategies, CFOs can identify new markets, expand their customer base, and explore new business models.

For instance, by analysing customer data, CFOs can identify untapped market segments and develop targeted marketing campaigns. By leveraging social media platforms and digital advertising, CFOs can reach a wider audience and generate more leads. Additionally, by adopting e-commerce platforms, CFOs can expand their reach beyond traditional brick-and-mortar stores and tap into the growing online marketplace.

While digital strategies offer immense potential for business growth, CFOs must carefully consider a few key factors to ensure successful implementation.

Firstly, CFOs must assess the organisation's readiness for digital transformation. This involves evaluating the current technological infrastructure, identifying any skill gaps, and developing a comprehensive roadmap for implementation.

Secondly, CFOs must prioritise security and data privacy. With the increasing digitisation of financial processes, cybersecurity has become a critical concern. CFOs must invest in robust security measures to protect sensitive financial information and ensure compliance with data privacy regulations.

Lastly, CFOs must foster a culture of innovation and continuous learning within the organisation. Digital transformation requires a mindset shift and a willingness to embrace change. CFOs must encourage employees to embrace new technologies and provide training and support to facilitate the adoption of digital strategies.

To understand the practical application of aligning digital strategies with business outcomes in finance, let's explore some successful case studies:

Standard Chartered and their Digital Transformation Journey: Standard Chartered, a leading financial institution, embarked on a digital transformation journey to enhance its customer experience and drive business outcomes.[1] By leveraging data analytics, the bank gained insights into customer preferences, enabling them to personalise their offerings and improve cross-selling opportunities. They also implemented automation and AI technologies to streamline their back-office operations, reducing costs and improving operational efficiency. As a result, Standard Chartered experienced significant growth in customer acquisition, retention, and overall profitability.

AXA Insurance's Digital Marketing Success: AXA Insurance, a leading insurance provider, embraced digital marketing to drive business outcomes.[2] Through targeted online advertising campaigns and content marketing strategies, AXA Insurance increased its brand visibility and generated high-quality leads. By leveraging customer data, they personalised their marketing messages and provided tailored insurance solutions, resulting in higher conversion rates and improved business outcomes.

Barclays Bank: Barclays Bank set out in their Strategic Report that they have embarked on a digital transformation journey with the goal of improving customer experience and expanding their digital journeys.[3] They implemented a digital banking platform that allowed customers to access their accounts, make payments, and apply for loans online. They are introducing digital tools to the Barclays app to provide new products for their customers, improve the overall experience, and enable individuals to manage their finances better. This not only enhanced customer satisfaction but also increased operational efficiency, as manual processes were automated.

Allianz Insurance: Allianz Insurance leveraged digital strategies to streamline their claims processing system. Allianz's Global Digital Factory launched a digital delivery hub to achieve change through digital projects, such as developing claims solutions, across its international operations.[4] By implementing automation in its claims management, they were able to reduce the time it took to process claims, resulting in improved customer satisfaction and reduced operational costs. Additionally, this provided real-time insights into claims data, enabling Allianz to identify patterns and detect potential fraud.

Tools and Technologies can be vital in supporting digital finance strategies. To support the implementation of digital finance strategies, CFOs can leverage a range of tools and technologies. Here are a few examples:

Cloud-Based Financial Management Systems: These systems allow CFOs to access financial data from anywhere, at any time, and facilitate collaboration among team members.

Data Analytics Platforms: These platforms enable CFOs to analyse large datasets and gain insights into financial performance, customer behaviour, and market trends.

Blockchain Technology: CFOs can leverage blockchain technology to enhance the security and transparency of financial transactions, reducing the risk of fraud.

Robotic Process Automation (RPA): RPA can automate repetitive and manual financial processes, freeing up time for CFOs to focus on strategic initiatives.

The future of digital finance promises even greater opportunities for business growth. As technologies continue to evolve, CFOs can expect to see advancements in areas such as artificial intelligence, machine learning, and predictive analytics.

These advancements will enable CFOs to make even more accurate financial forecasts, identify emerging risks, and develop proactive strategies to mitigate them. Additionally, as customer expectations continue to evolve, CFOs will need to leverage digital strategies to deliver personalised financial services and enhance the overall customer experience.

The role of CFOs in driving business growth has evolved significantly in the digital age. By embracing digital strategies, CFOs can unlock new opportunities, create innovative products and services, and position their organisations for success.

To effectively drive business growth through digital strategies, CFOs must align these strategies with broader business outcomes, leverage technology to create new products and services, and identify opportunities for growth through digital transformation. By carefully considering key factors and leveraging the right tools and technologies, CFOs can navigate the digital landscape and drive business growth with confidence.

It is clear that the future of finance lies in digital strategies. CFOs who embrace these strategies and adapt to the changing landscape will be well-positioned to drive business growth and deliver value to their organisations.

NOTES

1. https://www.sc.com/en/story/driving-digital-transformation-at-standard-chartered-rakeshs-story/
2. https://www.axaconnect.co.uk/broker-business/digital-marketing-the-benefits-of-using-digital-tools/
3. https://home.barclays/content/dam/home-barclays/documents/investor-relations/reports-and-events/annual-reports/2022/AR/Barclays-PLC-Strategic-Report-2022.pdf
4. https://www.mckinsey.com/industries/financial-services/our-insights/claims-in-the-digital-age

CHAPTER 31

Strategic Planning

Optimising Budgeting and Forecasting

Finance transformation is a critical process that helps organisations improve their financial operations, maximise efficiency, and drive growth. It involves adopting new technologies, streamlining processes, and rethinking financial strategies. But rigid budgets can block the ability of business leaders to pivot and capture new opportunities as they develop. Strategic planning plays a pivotal role in finance transformation, enabling businesses to optimise their budgeting and forecasting efforts. By aligning financial goals with overall business objectives, organisations can make informed decisions, adapt to market changes, and achieve long-term success.

Strategic planning serves as a roadmap for finance transformation. It involves setting clear objectives, defining strategies, and outlining action plans to achieve financial goals. By conducting a thorough analysis of the organisation's current financial state and future aspirations, strategic planning helps identify areas for improvement and prioritise initiatives.

Ahmad (2024) argues strategic planning ensures that budgeting and forecasting efforts are aligned with the overall strategic direction of the company, enabling finance teams to make informed decisions and allocate resources effectively (Ahmad, 2024).

Implementing *strategic planning in budgeting and forecasting* offers numerous benefits to organisations. Firstly, it enhances accuracy and reliability by providing a structured framework for financial analysis and decision-making. Strategic planning allows finance teams to consider various factors such as market trends, customer demands, and internal capabilities, resulting in more accurate budgeting and forecasting. This, in turn, leads to better resource allocation, risk management, and financial performance.

Secondly, strategic planning improves collaboration and communication within the organisation. By involving key stakeholders in the planning process, finance teams can gain valuable insights, align goals, and foster a shared understanding of financial objectives. This collaboration ensures that budgeting and forecasting efforts are based on a holistic view of the business, reducing silos and promoting a culture of transparency and accountability.

DOI: 10.1201/9781003514503-31

Vunjak et al. (2012) contend strategic planning enables agility and adaptability in a rapidly changing business environment (Vunjak et al., 2012). By continuously monitoring and evaluating financial performance, organisations can quickly identify deviations from the plan and make necessary adjustments. This proactive approach helps businesses stay ahead of market dynamics, seize opportunities, and mitigate risks effectively.

To optimise budgeting and forecasting efforts, organisations need to establish an *effective strategic planning process*. This process should include the following key components:

Mission and Vision: Clearly define the organisation's mission and vision statements, which provide a sense of purpose and direction for financial strategies.

Environmental Analysis: Conduct a thorough analysis of the internal and external factors that impact the organisation's financial performance. This includes assessing market trends, competitor analysis, and SWOT (Strengths, Weaknesses, Opportunities, and Threats) analysis.

Goal Setting: Set specific, measurable, achievable, relevant, and time-bound (SMART) financial goals that are aligned with the organisation's overall objectives. These goals should be challenging yet attainable to drive performance.

Strategy Development: Develop strategies and action plans to achieve the financial goals. This involves identifying key initiatives, allocating resources, and assigning responsibilities to ensure effective execution.

Performance Measurement: Establish key performance indicators (KPIs) and metrics to monitor and evaluate financial performance. Regularly track progress against the defined targets and make data-driven decisions to optimise budgeting and forecasting efforts.

Review and Continuous Improvement: Conduct regular reviews of the strategic plan, assess its effectiveness, and make necessary adjustments. Embrace a culture of continuous improvement to enhance financial processes and adapt to changing business conditions.

To *optimise budgeting and forecasting efforts*, organisations can follow these tips:

Leverage Technology: Utilise advanced financial planning and analysis (FP&A) tools and technologies to automate repetitive tasks, improve data accuracy, and enhance forecasting capabilities.

Collaborate Cross-Functionally: Engage with stakeholders from different departments to gather insights, align goals, and improve the accuracy of budgeting and forecasting.

Adopt Rolling Forecasts: Move away from traditional annual budgets and embrace rolling forecasts that provide a more dynamic and agile approach to financial planning.

Embrace Scenario Planning: Develop multiple scenarios to assess the impact of different market conditions on financial performance. This helps organisations prepare for uncertainties and make informed decisions.

Invest in Training and Development: Ensure that finance teams have the necessary skills and knowledge to effectively utilise financial planning tools and technologies. Continuous training and development programmes can enhance their capabilities and drive better budgeting and forecasting outcomes.

Promote a Data-Driven Culture: Encourage the use of data and analytics in decision-making processes. Foster a culture that values evidence-based insights and encourages financial teams to leverage data for accurate budgeting and forecasting.

Finance transformation initiatives often face several challenges, including resistance to change, legacy systems, and limited resources. However, strategic planning can help overcome these challenges by providing a structured approach to navigate through transformational journeys.

Firstly, strategic planning helps address resistance to change by involving key stakeholders in the planning process. By creating a shared vision and fostering open communication, individuals become more invested in the transformation initiatives, leading to higher acceptance and engagement.

Secondly, strategic planning enables organisations to identify and address the limitations posed by legacy systems. By conducting a thorough analysis of existing technologies and processes, finance teams can develop strategies to modernise systems and streamline operations, resulting in improved efficiency and effectiveness.

Lastly, strategic planning helps optimise resource allocation by identifying areas of investment that provide the highest returns. By evaluating the cost-benefit ratio of different initiatives, organisations can prioritise projects that align with strategic objectives and maximise the utilisation of limited resources.

Tools and technologies are available to enhance budgeting and forecasting processes. These include:

Enterprise Resource Planning (ERP) Systems: ERP systems integrate various financial functions, enabling real-time data analysis, enhanced reporting capabilities, and better budgeting and forecasting accuracy.

FP&A Software: FP&A software automates budgeting and forecasting processes, improves data accuracy, and provides advanced analytics capabilities for better decision-making.

Cloud-Based Financial Management Systems: Cloud-based systems offer scalability, accessibility, and collaboration features that enhance budgeting and forecasting efforts. They also provide real-time visibility into financial data, enabling faster and more accurate decision-making.

Data Visualisation Tools: Data visualisation tools transform complex financial data into visually appealing and easy-to-understand charts and graphs. These tools enhance communication and aid in decision-making processes.

Artificial Intelligence (AI) and Machine Learning (ML): AI and ML technologies can analyse vast amounts of financial data, identify patterns, and provide predictive insights for budgeting and forecasting purposes. These technologies enable organisations to make data-driven decisions and improve forecasting accuracy.

Several organisations have successfully implemented finance transformation initiatives by leveraging strategic planning. Let's explore two case studies which demonstrate the power of strategic planning in finance transformation and its impact on budgeting and forecasting efforts:

Alten Group: Alten Group, a technology consulting company, faced challenges in aligning financial goals with overall business objectives. By implementing a strategic planning process, they were able to streamline budgeting and forecasting efforts. They implemented new cloud finance systems, ERP, and people platforms. This helped improve standardisation of processes, the same ways of working, and alignment. It provided a common and integrated financial group reporting.[1] As a result, Alten Group achieved improved financial visibility, enhanced collaboration between finance and other areas, improved centralisation, efficiency, and improved resource allocation.

WSP: Williams Sale Partnership (WSP), a professional services provider, experienced limitations with legacy systems that hindered their budgeting and forecasting processes. Through strategic planning, they identified the need for technology modernisation and developed a roadmap to upgrade their financial systems.[2] By leveraging cloud-based financial management systems and FP&A software, WSP improved data accuracy, enhanced reporting capabilities, and achieved real-time visibility into financial data. This enabled them to make more informed decisions, optimise resource allocation, and drive better financial performance.

To effectively implement strategic planning in finance transformation, organisations should consider the following best practices:

Leadership Commitment: Obtain buy-in from top management and ensure they actively participate in the strategic planning process. Leadership commitment sets the tone for the entire organisation and enhances the chances of successful implementation.

Engage Cross-Functional Teams: Involve key stakeholders from various departments to gather diverse perspectives and ensure alignment with overall business objectives. This collaboration fosters a sense of ownership and collective responsibility for financial outcomes.

Regular Communication and Feedback: Maintain open lines of communication throughout the strategic planning process. Seek feedback from stakeholders, address concerns, and provide updates on progress. This transparency enhances engagement and promotes a culture of continuous improvement.

Monitor and Track Progress: Establish a robust performance measurement system to monitor progress against defined targets. Regularly review KPIs and make data-driven decisions to optimise budgeting and forecasting efforts.

Flexibility and Adaptability: Recognise that strategic plans may require adjustments as market conditions and business dynamics evolve. Embrace a flexible mindset and be open to incorporating changes to ensure the relevance and effectiveness of the strategic planning process.

Strategic planning plays a crucial role in finance transformation, enabling organisations to optimise budgeting and forecasting efforts. By aligning financial goals with overall business objectives, strategic planning enhances accuracy, improves collaboration, and enables agility in a rapidly changing business environment. Key components of an effective strategic planning process include mission and vision statements, environmental analysis, goal setting, strategy development, performance measurement, and continuous improvement.

Organisations can optimise budgeting and forecasting efforts by leveraging technology, promoting cross-functional collaboration, adopting rolling forecasts, embracing scenario planning, investing in training and development, and promoting a data-driven culture.

Strategic planning helps overcome common challenges in finance transformation, such as resistance to change, legacy systems, and limited resources. Tools and technologies like ERP systems, FP&A software, cloud-based financial management systems, data visualisation tools, AI, and ML enhance budgeting and forecasting processes.

Real-world case studies demonstrate the success of strategic planning in finance transformation. To implement strategic planning effectively, organisations should follow best practices such as leadership commitment, cross-functional engagement, regular communication and feedback, monitoring and tracking progress, and flexibility and adaptability.

By leveraging the power of strategic planning, organisations can optimise their budgeting and forecasting efforts, drive financial performance, and achieve long-term success in finance transformation.

NOTES

1. https://www.unit4.com/sites/default/files/documents/Unit4-Alten-infographic-CT200710INT.pdf
2. https://www.unit4.com/sites/default/files/library-files/2023-02/Unit4-all-fpa-2020-cs-wsp-CT200414INT.pdf

CHAPTER 32

Finance Specialisms

As the business landscape continues to evolve and become more complex, finance departments are faced with the challenge of keeping up with the rapid pace of change. In order to remain competitive and meet the demands of the modern business world, organisations are turning to finance transformation.

Charkha (2023) contends that finance processes require specialist subject matter experts and that digital transformation is putting increasing pressure to establish finance professionals as effective business partners, taking on digital technological skills, and that the finance function is incorporating risk-intelligent planning models and deep data-driven business knowledge and foresight (Charkha, 2023). Finance processes are even more niche and critical whilst finance transformation is undertaken. Chief Finance Officer (CFO) must support organisations to improve their capabilities in these specialisms to ensure they enable transformation efforts.

- *Accounting:* From technical accounting issues such as Generally Accepted Accounting Principles (GAAP) and International Financial Reporting Standards (IFRS) to risk and compliance programmes and controls, we bring a depth of accounting experience and insight to each finance transformation initiative.

- *Tax:* How does your organisation's tax strategy – including legal entities and tax structures – align with its operating model? Working with clients, we help provide tight alignment by implementing solutions that are both scalable and sustainable.

- *Treasury:* Improving the end-to-end cycle of the treasury isn't just a matter of addressing the individual dimensions of processes, people, or technology – it requires addressing them in tandem. That's where we can make connections for clients that others might miss.

- *Investor Relations:* Keeping investors abreast of and engaged with finance goals is a responsibility that is only gaining in importance – and yet it is still too often overlooked. We help clients execute and communicate a cohesive value proposition to both internal and external stakeholders.

DOI: 10.1201/9781003514503-32

- *M&A Finance:* In the face of a merger or acquisition, will the finance organisation move to a joint function, or will it divest functions? We have deep experience with both and will work with you to choose the desired path for your organisation to keep risk and costs in check.

As digital transformation in finance is a strategic initiative, it involves reimagining and reengineering financial processes, systems, and technologies to drive efficiency, improve decision-making, and enhance the overall performance of the accounting and treasury functions.

Accounting and Treasury are vital functions to transform. Finance transformation plays a crucial role in the accounting and treasury functions of organisations. Traditionally, these departments have been seen as cost centres, focused on transactional processing and compliance. However, with the advent of finance transformation, their role has shifted to that of a strategic partner. Finance transformation enables accounting and treasury professionals to move beyond their traditional responsibilities and become trusted advisors to the business.

One of the key aspects of finance transformation in accounting and treasury is the automation of manual processes. By leveraging technology, organisations can streamline their financial operations, reduce errors, and improve the accuracy and timeliness of financial reporting. This not only saves time and resources but also allows finance professionals to focus on more value-added activities, such as financial analysis and strategic planning.

Accounting and Treasury are key areas for transformation. The benefits of finance transformation in accounting and treasury are numerous and far-reaching. Firstly, it leads to increased operational efficiency. By automating manual processes and implementing streamlined workflows, organisations can reduce the time and effort required to complete financial tasks. This frees up valuable resources that can be redirected towards more strategic initiatives.

Secondly, finance transformation improves data accuracy and integrity. By digitising financial processes and implementing robust control mechanisms, organisations can minimise the risk of errors and fraud. This not only enhances the reliability of financial information but also allows for better decision-making based on accurate and up-to-date data.

Furthermore, finance transformation enables organisations to gain real-time visibility into their financial performance. By leveraging advanced analytics and reporting tools, finance professionals can access timely and actionable insights, allowing them to make informed decisions and drive business growth.

Several key trends are driving finance transformation in accounting and treasury. One of the most prominent trends is the adoption of cloud-based solutions. Cloud computing offers numerous benefits, including scalability, cost-effectiveness, and enhanced collaboration. By migrating their financial systems to the cloud, organisations can access their data from anywhere, at any time, and streamline their financial processes.

Another trend is the increasing focus on data analytics and artificial intelligence (AI). With the vast amount of data available to organisations, leveraging analytics and

AI can provide valuable insights and improve decision-making. By harnessing the power of these technologies, finance professionals can identify patterns, trends, and anomalies in financial data, enabling them to make data-driven decisions and drive business performance.

Additionally, there is a growing emphasis on risk management and compliance. In an increasingly regulated business environment, organisations need to ensure that their financial processes and systems comply with relevant regulations. Finance transformation enables organisations to implement robust control mechanisms and automated compliance checks, reducing the risk of non-compliance and potential penalties.

Implementing finance transformation in accounting and treasury requires careful planning and execution. The first step is to assess the current state of the finance function and identify areas for improvement. This involves conducting a thorough review of existing processes, systems, and technologies and identifying pain points and inefficiencies.

Once the areas for improvement have been identified, organisations can develop a finance transformation roadmap. This roadmap outlines the desired future state of the finance function and the steps required to get there. It includes a detailed plan for implementing new processes, systems, and technologies, as well as a timeline and resource allocation.

During the implementation phase, organisations need to ensure effective change management. This involves communicating the vision and goals of finance transformation to all stakeholders and providing the necessary training and support. It is crucial to address any resistance to change and ensure that employees understand the benefits of finance transformation and are fully engaged in the process.

While finance transformation offers numerous benefits, it also presents challenges and considerations for accounting and treasury professionals. One of the key challenges is the complexity of implementing new technologies and systems. Finance professionals may lack the necessary technical expertise to fully leverage these tools, and organisations need to invest in training and support to ensure successful implementation.

Another consideration is the potential impact on the workforce. As manual processes are automated, there may be concerns about job security and the need for upskilling.

Organisations need to proactively address these concerns and provide opportunities for professional development and career advancement.

Furthermore, finance transformation requires a significant investment of time, resources, and capital. Organisations need to carefully evaluate the costs and benefits of finance transformation and ensure that the expected return on investment justifies the expenditure.

The future of finance transformation in accounting and treasury is promising. As technology continues to advance, organisations can expect further automation and digitisation of financial processes. The use of artificial intelligence and machine learning algorithms will enable organisations to gain even deeper insights from their financial data and make more accurate predictions.

Furthermore, the integration of finance and other business functions will become more seamless. Finance transformation will enable organisations to break down silos and promote cross-functional collaboration, leading to more effective decision-making and improved business outcomes.

Additionally, there will be a greater focus on sustainability and environmental, social, and governance (ESG) reporting. Finance transformation will enable organisations to capture and analyse ESG data, allowing them to demonstrate their commitment to sustainability and responsible business practices.

There are several tools and technologies that can facilitate finance transformation in accounting and treasury. One such tool is enterprise resource planning (ERP) software. ERP systems integrate various financial processes, such as accounting, budgeting, and reporting, into a single platform, providing organisations with real-time visibility and control over their financial operations.

Another technology is robotic process automation (RPA). RPA involves the use of software robots to automate repetitive and rule-based tasks, such as data entry and reconciliation. By leveraging RPA, organisations can improve efficiency, accuracy, and compliance in their financial processes.

Additionally, advanced analytics and reporting tools, such as data visualisation and predictive analytics software, can provide finance professionals with actionable insights and enable them to make data-driven decisions.

Several organisations have successfully implemented finance transformation in their accounting and treasury functions. One such case study is *Johnson & Johnson*, a multinational corporation in the manufacturing industry. By leveraging cloud-based ERP software and advanced analytics tools, Johnson & Johnson was able to streamline its financial processes, improve data accuracy, and gain real-time visibility into its financial performance. As a result, the company was able to make more informed decisions, reduce costs, and drive business growth.

Another case study is *Transamerica Corporation*, a financial services firm. Transamerica Corporation implemented RPA in its accounting and treasury functions, automating manual processes and improving efficiency and accuracy. This enabled the company to redirect its finance professionals towards more value-added activities, such as financial analysis and strategic planning, leading to improved business performance and customer satisfaction.

Finance transformation is revolutionising the accounting and treasury functions of organisations. By reimagining and reengineering financial processes, systems, and technologies, organisations can drive efficiency, improve decision-making, and enhance the overall performance of their finance departments. The benefits of finance transformation are numerous and include increased operational efficiency, improved data accuracy, and real-time visibility into financial performance. However, implementing finance transformation presents challenges and considerations, such as the complexity of new technologies and the potential impact on the workforce. Nevertheless, the future of finance transformation is promising, with further advancements in technology and increased integration with other business functions. By leveraging tools and technologies such as ERP software, RPA, and advanced analytics, organisations can successfully navigate the future and achieve finance transformation in accounting and treasury.

As a financial professional, I understand the importance of staying ahead in a rapidly evolving business landscape. One of the key ways to navigate change is through finance transformation. CFOs must consider how finance transformation can revolutionise tax, investor relations, and M&A finance. They must delve into the impact of finance transformation on tax, how it can be leveraged for investor relations, and how it streamlines M&A finance. Additionally, they must consider key considerations for implementing finance transformation, look at successful case studies, address potential challenges, and examine the future of finance transformation in these areas.

Finance transformation is a strategic initiative that aims to improve the efficiency and effectiveness of financial processes within an organisation. It involves re-evaluating and restructuring finance operations, systems, and technologies to drive business growth, enhance decision-making, and mitigate risk. Finance transformation goes beyond mere cost-cutting measures; it aims to optimise financial processes, streamline operations, and enable better data-driven decision-making.

Tax is a critical aspect of any organisation's financial management. Finance transformation can have a significant impact on tax processes and outcomes. By implementing finance transformation initiatives, companies can enhance tax compliance, improve reporting accuracy, and increase efficiency in tax planning and forecasting.

One important way finance transformation revolutionises tax is through automation. By leveraging advanced technologies such as artificial intelligence and machine learning, finance transformation enables the automation of routine tax processes, reducing manual errors and saving time. This allows tax professionals to focus on more strategic and value-added tasks, such as tax planning and analysis.

Furthermore, finance transformation facilitates better visibility and control over tax data. Integrated finance systems provide a centralised repository for tax-related information, enabling real-time access to data and improving collaboration between finance and tax teams. This enhances data accuracy, reduces duplication, and ensures compliance with changing tax regulations.

Investor relations play a crucial role in maintaining strong relationships with shareholders and attracting new investors. Finance transformation can be a game-changer in this area by providing better financial insights, enhancing communication, and improving investor confidence.

One of the key advantages of finance transformation for investor relations is the ability to generate timely and accurate financial reports. By streamlining financial processes and implementing robust reporting systems, finance transformation enables companies to produce high-quality financial statements and investor presentations. This allows for better transparency and provides investors with the information they need to make informed decisions.

Additionally, finance transformation can help companies analyse and interpret financial data more effectively. With advanced analytics tools, finance professionals can gain deeper insights into key performance indicators, financial trends, and market conditions. This enables them to communicate financial information more confidently, answer investor queries more comprehensively, and make data-driven recommendations to management.

Mergers and acquisitions (M&A) finance can be complex and time-consuming. Finance transformation can simplify and streamline the M&A process, making it more efficient and cost-effective.

One way finance transformation revolutionises M&A finance is through improved due diligence. By digitising and automating due diligence processes, finance transformation enables a faster and more accurate assessment of financial risks and opportunities. This allows companies to make informed decisions and negotiate favourable terms during M&A transactions.

Furthermore, finance transformation facilitates the seamless integration of acquired companies into the existing finance systems and processes. By standardising financial reporting, consolidating data, and implementing common financial platforms, finance transformation enables smooth post-merger integration. This not only reduces duplication and improves efficiency but also enhances financial control and visibility across the merged entities.

Implementing finance transformation requires careful planning and consideration. Before embarking on a finance transformation journey, organisations need to evaluate their current financial processes, identify pain points and opportunities for improvement, and define their strategic objectives.

One key consideration is selecting the right finance transformation tools and technologies. It is essential to assess the organisation's needs and choose solutions that align with its goals and budget. This may involve implementing ERP systems, adopting advanced analytics tools, or leveraging RPA technologies.

Another crucial aspect is change management. Finance transformation initiatives often involve significant organisational change, and it is important to get buy-in from key stakeholders and ensure proper communication and training. This helps employees understand the benefits of finance transformation, adapt to new processes, and embrace the changes positively.

To illustrate the power of finance transformation in tax, investor relations, and M&A finance, let's look at a few successful case studies.

Expedia Group: Expedia Group, a multinational company, implemented finance transformation initiatives to streamline its tax processes. By leveraging automation and advanced analytics, XYZ Corporation reduced the time required for tax compliance, improved accuracy in tax reporting, and enhanced tax planning capabilities. As a result, the company achieved significant cost savings and better compliance with tax regulations.[1]

JP Morgan Chase: JP Morgan Chase, a leading financial institution, embraced finance transformation to enhance its investor relations. By implementing robust reporting systems and leveraging data analytics, the bank improved the quality and timeliness of its financial disclosures. This led to increased investor confidence, improved shareholder relationships, and ultimately higher market valuation.

Deloitte: Deloitte utilised finance transformation to streamline its M&A finance processes. By digitising due diligence, standardising financial reporting, and integrating acquired companies into its finance systems, Deloitte reduced the time and cost involved in M&A transactions. This allowed the company to pursue more strategic acquisitions and achieve faster post-merger integration.

While finance transformation offers immense benefits, it is not without challenges. Organisations may face resistance to change, a lack of expertise in implementing new technologies, and the need for significant investments in infrastructure and training.

One key challenge is data integration. Finance transformation often involves consolidating data from multiple sources, which can be complex and time-consuming. Organisations need to ensure data accuracy, data security, and seamless integration across different systems and processes.

Another challenge is managing organisational change. Finance transformation requires organisations to adopt new processes, technologies, and ways of working. Resistance to change may arise from employees who are comfortable with existing systems or fear job displacement. Proper change management strategies, including communication, training, and employee engagement, are essential to overcome these challenges.

Partnering with finance transformation consultants can be a key lifeline. To help ensure successful finance transformation, organisations can consider partnering with finance transformation consultants. These professionals bring expertise in finance processes, technology implementation, and change management. They can help organisations define their finance transformation roadmap, select the right tools and technologies, and navigate potential challenges.

Finance transformation consultants can also provide valuable insights and best practices based on their experience in implementing similar initiatives in other organisations. They can facilitate knowledge transfer, train employees, and provide ongoing support during the finance transformation journey.

Looking ahead, the future of finance transformation in tax, investor relations, and M&A finance is promising. As technology continues to advance, finance professionals can leverage emerging tools such as blockchain, predictive analytics, and RPA to further enhance financial processes and decision-making.

In tax, advanced analytics and machine learning algorithms can enable more accurate tax planning and forecasting. Blockchain technology can revolutionise tax compliance by providing a secure and transparent platform for recording and verifying transactions. Additionally, RPA can automate routine tax processes, freeing up resources for more strategic tax initiatives.

For investor relations, predictive analytics can help identify investor preferences and anticipate market trends, enabling more targeted and effective communication. Virtual reality and augmented reality technologies can provide immersive experiences for investors, enhancing their understanding of financial data and performance.

In M&A finance, artificial intelligence can streamline due diligence processes by analysing vast amounts of data and identifying potential risks and opportunities. Blockchain can provide secure and transparent transaction records, simplifying post-merger integration.

Additionally, RPA can automate repetitive tasks in the M&A process, improving efficiency and reducing costs.

Finance transformation has the potential to revolutionise tax, investor relations, and M&A finance. By embracing finance transformation initiatives, organisations can enhance tax compliance, improve investor relations, and streamline M&A processes. However, implementing finance transformation requires careful planning, change management, and the right tools and technologies. By partnering with finance transformation consultants and staying abreast of emerging technologies, organisations can navigate change and stay ahead in the evolving business landscape.

NOTE

1. https://www.pwc.com/gx/en/tax/publications/assets/enhancing-tax-process-managemnt-and-controls.pdf

CHAPTER 33

How Transformation Can Revolutionise Performance Management

FINANCE TRANSFORMATION REFERS TO the process of reimagining and redefining the role of finance within an organisation. It involves leveraging technology, streamlining processes, and adopting best practices to drive efficiency, effectiveness, and value creation. In today's rapidly changing business landscape, finance transformation has become a critical strategic initiative for organisations seeking to stay competitive and achieve sustainable growth.

Performance management is a crucial aspect of any organisation's success. It involves setting clear goals, monitoring progress, providing feedback, and aligning individual and team performance with organisational objectives. Effective performance management ensures that employees are engaged, motivated, and focused on achieving results that contribute to the overall success of the organisation.

Hakizimana et al. (2023) argue digital transformation in finance is vital as traditional performance management practices often fall short in delivering the desired outcomes (Hakizimana et al., 2023). They are often characterised by annual or semi-annual performance reviews that focus on past performance rather than future potential. These reviews can be subjective, time-consuming, and demotivating for employees. Additionally, traditional performance management systems may lack the flexibility to adapt to changing business needs and may not provide timely and actionable insights for decision-making.

Finance transformation can play a pivotal role in revolutionising performance management within an organisation. By leveraging technology and data analytics, finance teams can provide real-time and actionable insights into performance metrics. This enables managers to make informed decisions, identify areas for improvement, and drive performance at both the individual and organisational levels.

DOI: 10.1201/9781003514503-33

Finance transformation for performance management encompasses several key components. Firstly, it involves the adoption of a performance management framework that aligns individual goals with organisational objectives. This framework should provide clear expectations, regular feedback, and opportunities for skill development and growth.

Secondly, finance transformation entails the implementation of robust performance measurement systems that capture relevant data and metrics. This enables managers to track progress, identify trends, and make data-driven decisions. Lastly, finance transformation requires the integration of technology solutions that automate manual processes, enhance data accuracy, and provide real-time reporting and analytics.

Implementing finance transformation in performance management involves a systematic approach. Organisations should start by assessing their current performance management practices and identifying areas for improvement. This can be done through employee surveys, focus groups, and benchmarking against industry best practices. Once the areas for improvement are identified, organisations can design and implement a finance transformation roadmap that outlines the necessary changes, timelines, and resource requirements. It is crucial to involve key stakeholders, such as finance and HR teams, in the implementation process to ensure buy-in and alignment with organisational goals.

Finance transformation in performance management can yield several benefits for organisations. Firstly, it enables organisations to make better-informed decisions by providing real-time and accurate performance data. This improves agility and responsiveness to changing market conditions. Secondly, finance transformation enhances employee engagement and motivation by providing timely feedback and recognition. This leads to increased productivity and the retention of top talent. Lastly, finance transformation drives efficiency and cost savings by automating manual processes and eliminating redundant tasks. This enables finance teams to focus on value-added activities that contribute to the organisation's strategic objectives.

We can consider some case studies. Several organisations have successfully implemented finance transformation in performance management. One such example is *Google*, a multinational corporation in the technology sector. By adopting a performance management framework that aligns individual goals with organisational objectives, Google was able to improve employee engagement and productivity.[1] Google adopted the OKR performance framework in 1999 and has taken it to the next level by making suitable tweaks to the system to make all its functions more effective and its workforce more skilful. They implemented a robust performance measurement system that provided real-time insights into key performance indicators. As a result, Google achieved significant improvements in financial performance and gained a competitive edge in the market.

Another case study is *Wells Fargo*, a multinational financial firm. Through finance transformation, Wells Fargo implemented technology solutions that automated manual processes, such as performance data collection and reporting. This streamlined their performance management practices and enabled managers to make data-driven decisions. Wells Fargo also focused on providing regular feedback and recognition to employees, leading to improved morale and job satisfaction. As a result, they experienced higher employee retention rates and achieved greater operational efficiency.

Best practices are vital for consistency. To ensure successful implementation of finance transformation in performance management, organisations should adhere to best practices. Firstly, it is essential to have strong leadership support and buy-in from key stakeholders. This ensures alignment with organisational goals and secures the necessary resources for implementation. Secondly, organisations should invest in robust technology solutions that are scalable, user-friendly, and integrate seamlessly with existing systems. This enables efficient data collection, analysis, and reporting. Lastly, organisations should prioritise change management and provide adequate training and support to employees. This fosters a culture of continuous improvement and ensures the adoption of new performance management practices.

Finance transformation has the potential to revolutionise performance management within organisations. By leveraging technology, streamlining processes, and adopting best practices, organisations can unlock success and achieve sustainable growth. Effective performance management, enabled by finance transformation, drives employee engagement, improves decision-making, and enhances operational efficiency. By implementing finance transformation in performance management, organisations can position themselves as industry leaders and stay ahead in today's dynamic business environment.

Finance transformation is a critical process for organisations aiming to stay competitive in today's rapidly changing business landscape. To effectively navigate this transformation, it is important to understand the role of performance management and how digital business cases can unlock its potential. CFOs must explore digital business cases, discuss the opportunities and challenges associated with their implementation, and consider the significance of return-based measures. We can consider case studies, outline steps to unlock the potential of finance transformation through performance management, and consider the best practices for implementation. By harnessing the power of finance transformation through performance management, organisations can achieve their strategic goals and drive sustainable growth.

Performance management plays a crucial role in finance transformation. It encompasses the processes, methodologies, and tools used to measure and improve an organisation's performance. By implementing effective performance management practices, organisations can align their financial goals with their overall business objectives. This alignment enables them to make informed decisions, optimise resource allocation, and drive operational efficiency.

In the context of finance transformation, performance management becomes even more critical. It provides the framework for assessing and improving the financial performance of an organisation during the transformation process. Through the use of key performance indicators (KPIs) and other performance measurement tools, organisations can monitor progress, identify areas for improvement, and take corrective actions as needed. Performance management also enables organisations to track the impact of transformation initiatives on financial outcomes and ensure that the desired results are achieved.

Digital business cases are an integral part of performance management in finance transformation. These cases leverage digital technologies and data analytics to improve the accuracy, timeliness, and relevance of financial information. By digitising financial processes and leveraging advanced analytics, organisations can gain deeper insights into their financial performance, identify patterns and trends, and make data-driven decisions.

One of the key advantages of digital business cases is their ability to provide real-time visibility into financial data. Traditional performance management processes often rely on manual data collection and reporting, which can be time-consuming and prone to errors. Digital business cases automate these processes, enabling organisations to access up-to-date financial information whenever they need it. This real-time visibility allows for faster decision-making and timely interventions to address any issues that may arise during the finance transformation journey.

Moreover, digital business cases enable organisations to leverage advanced analytics techniques such as predictive modelling and machine learning. These techniques can help identify patterns and correlations in financial data, uncover hidden insights, and generate accurate forecasts. By harnessing the power of these advanced analytics capabilities, organisations can make more accurate predictions about future financial performance and proactively address any potential risks or opportunities.

Implementing performance management digital business cases offers numerous opportunities for organisations undergoing finance transformation. These opportunities include improved data accuracy, enhanced decision-making capabilities, increased operational efficiency, and better alignment of financial goals with business objectives. However, along with these opportunities, organisations also face several challenges that need to be addressed.

One of the key challenges in implementing performance management digital business cases is data integration. Organisations often have multiple systems and sources of financial data, which may not be compatible with each other. Integrating these disparate data sources and ensuring data accuracy and consistency can be a complex and time-consuming process.

Organisations need to invest in robust data integration and governance frameworks to overcome these challenges and ensure the reliability and integrity of their financial data.

Another challenge is the need for skilled resources to manage and analyse digital business cases. Implementing digital technologies and advanced analytics requires specialised skills in areas such as data science, data engineering, and data visualisation. Organisations need to invest in training and development programmes to build the necessary capabilities within their finance teams or consider partnering with external experts to leverage their expertise.

Furthermore, organisations need to overcome resistance to change and cultural barriers to successfully implement performance management digital business cases. Finance transformation often requires a shift in mindset and ways of working, which may be met with resistance from employees. Organisations need to communicate the benefits of the transformation, provide training and support to employees, and create a culture of continuous improvement to drive successful implementation.

By addressing these challenges and leveraging the opportunities offered by performance management digital business cases, organisations can unlock the full potential of finance transformation and achieve sustainable growth.

Return-based measures play a significant role in finance transformation. These measures provide insights into the financial performance and efficiency of an organisation. By analysing return-based measures, organisations can evaluate the effectiveness of their strategies, assess the value generated from their investments, and identify areas for improvement.

One key return-based measure is return on investment (ROI). ROI measures the return generated from an investment relative to its cost. It provides an indication of the profitability and efficiency of an investment and helps organisations assess the value created through their finance transformation initiatives. By tracking ROI, organisations can identify high-performing investments and allocate resources accordingly.

Another return-based measure is return on equit (ROE). ROE measures the profitability of an organisation relative to its shareholders' equity. It indicates how effectively the organisation is utilising its equity to generate profits. ROE is a critical measure for investors and stakeholders, as it reflects the organisation's ability to generate returns on their investments.

In addition to ROI and ROE, organisations can use other return-based measures such as Return on Assets (ROA) and Return on Capital Employed (ROCE) to evaluate their financial performance. These measures provide a holistic view of the organisation's efficiency and profitability and help in identifying areas for improvement.

By leveraging return-based measures, organisations can monitor the financial impact of their finance transformation initiatives, evaluate their effectiveness, and make informed decisions to drive sustainable growth.

To understand the practical implications of performance management digital business cases in finance transformation, let's explore some real-world case studies.

BMW, a global manufacturing company, embarked on a finance transformation journey to streamline its financial processes and improve decision-making. By implementing a digital business case focused on real-time financial reporting and analytics, BMW gained visibility into their financial performance at various levels of the organisation. This visibility enabled them to identify cost-saving opportunities, optimise resource allocation, and enhance overall operational efficiency. As a result, BMW achieved a significant reduction in their operating costs and improved their profitability.

Transamerica Corporation, a financial services organisation, implemented a digital business case to automate their budgeting and forecasting processes. By leveraging advanced analytics and machine learning algorithms, Transamerica Corporation was able to generate accurate forecasts and optimise its budget allocations. This automation reduced the time and effort required for budgeting and forecasting, allowing the finance team to focus on more strategic activities. Transamerica Corporation experienced improved financial planning, better resource utilisation, and increased agility in responding to market changes.

These case studies demonstrate the transformative power of performance management digital business cases in finance transformation. By leveraging digital technologies and advanced analytics, organisations can achieve significant improvements in their financial performance, operational efficiency, and decision-making capabilities.

To unlock the potential of finance transformation through performance management, organisations can follow these essential steps:

Define Clear Objectives: Identify the specific goals and objectives of the finance transformation initiative. Align these objectives with the overall business strategy and ensure they are measurable and attainable.

Assess Current Performance Management Practices: Evaluate the existing performance management practices and identify areas for improvement. This assessment should include an analysis of the data collection and reporting processes, performance measurement frameworks, and the availability of relevant technologies and tools.

Design Performance Measurement Framework: Develop a comprehensive performance measurement framework that aligns with the organisation's goals and objectives. This framework should include relevant KPIs, targets, and benchmarks to measure financial performance accurately.

Implement Digital Business Cases: Identify the digital business cases that can enhance performance management in finance transformation. These cases may include automation of financial processes, digitisation of reporting and analytics, and the use of advanced analytics techniques.

Ensure Data Accuracy and Integration: Establish robust data integration and governance frameworks to ensure the accuracy and consistency of financial data. Implement data quality controls and validation processes to minimise errors and ensure data integrity.

Build Analytical Capabilities: Invest in training and development programs to build analytical capabilities within the finance team. This includes providing training on data analytics tools, techniques, and methodologies, as well as promoting a data-driven culture within the organisation.

Monitor and Evaluate Performance: Continuously monitor and evaluate the performance of the finance transformation initiatives. Regularly review the performance measurement framework, assess the accuracy and relevance of the KPIs, and make adjustments as needed.

Drive Continuous Improvement: Foster a culture of continuous improvement within the organisation. Encourage feedback and suggestions from employees, implement best practices, and continuously seek opportunities to enhance performance management practices.

By following these steps, organisations can unlock the full potential of finance transformation through performance management and drive sustainable growth.

In the era of digital transformation, several tools and technologies are available to support performance management in finance transformation. These tools enable organisations to automate financial processes, leverage advanced analytics, and improve decision-making capabilities. Let's explore some of the key tools and technologies used in performance management:

Enterprise Performance Management (EPM) Systems: EPM systems provide a comprehensive platform for managing financial performance. These systems integrate financial planning, budgeting, forecasting, and reporting processes, enabling organisations to streamline their performance management activities. EPM systems also provide advanced analytics capabilities, allowing for data-driven insights and decision-making.

Data Visualisation Tools: Data visualisation tools help organisations transform complex financial data into meaningful visual representations. These tools enable users to create interactive dashboards, charts, and graphs, making it easier to understand and analyse financial information. Data visualisation tools enhance communication and facilitate data-driven decision-making by presenting financial data in a visually appealing and easy-to-understand format.

Advanced Analytics Platforms: Advanced analytics platforms, such as predictive analytics and machine learning, enable organisations to uncover hidden insights and make accurate predictions about future financial performance. These platforms leverage algorithms and statistical models to analyse historical financial data, identify patterns and trends, and generate accurate forecasts. By leveraging advanced analytics, organisations can make data-driven decisions and optimise their financial performance.

Cloud-Based Solutions: Cloud-based solutions offer several advantages for performance management in finance transformation. These solutions provide scalability, flexibility, and cost-effectiveness, allowing organisations to leverage advanced technologies without significant upfront investments. Cloud-based solutions also facilitate collaboration and data sharing across geographically dispersed teams, enabling real-time access to financial information.

Artificial Intelligence (AI) and Robotic Process Automation (RPA): AI and RPA technologies automate repetitive financial processes, such as data entry and reconciliation, enabling organisations to improve efficiency and accuracy. These technologies can be integrated with existing financial systems to streamline workflows, reduce manual errors, and free up resources for more value-added activities.

By leveraging these tools and technologies, organisations can enhance their performance management capabilities, drive operational efficiency, and make data-driven decisions to support their finance transformation initiatives.

Implementing performance management digital business cases requires careful planning and execution. To ensure successful implementation, organisations should consider the following best practices:

Define Clear Objectives and Scope: Clearly define the objectives and scope of the digital business cases. Align these objectives with the overall finance transformation goals and ensure they are specific, measurable, attainable, relevant, and time-bound.

Engage Stakeholders: Involve key stakeholders, including finance leaders, business executives, and IT professionals, in the design and implementation of digital business cases. Engaging stakeholders from the outset ensures their buy-in and support, increasing the chances of successful implementation.

Prioritise Data Integration and Governance: Invest in robust data integration and governance frameworks to ensure the accuracy and consistency of financial data. Establish data quality controls, validation processes, and data governance policies to minimise errors and ensure data integrity.

Leverage Change Management Practices: Implement change management practices to address resistance to change and cultural barriers. Communicate the benefits of the digital business cases, provide training and support to employees, and create a culture of continuous improvement to drive successful implementation.

Start Small and Scale Up: Begin implementation with a pilot project or a smaller-scale initiative to test the effectiveness of the digital business case. Once the pilot project demonstrates positive results, gradually scale up the implementation to other areas of the organisation.

Monitor and Evaluate Performance: Continuously monitor and evaluate the performance of the digital business cases. Regularly review the KPIs, assess the accuracy and relevance of the data, and make adjustments as needed. Monitor the impact of digital business cases on financial outcomes and ensure that the desired results are achieved.

Promote Collaboration and Knowledge Sharing: Foster collaboration and knowledge sharing among finance teams and other stakeholders. Encourage cross-functional collaboration, facilitate information sharing, and promote a data-driven culture within the organisation.

By following these best practices, organisations can increase the likelihood of successful implementation of performance management digital business cases and drive positive outcomes in their finance transformation journey.

Finance transformation is a critical process for organisations aiming to stay competitive in today's business landscape. By effectively harnessing the power of performance management and leveraging digital business cases, organisations can unlock the full potential of finance transformation. Performance management provides the framework for assessing and improving financial performance during the transformation process. Digital business cases leverage advanced technologies and data analytics to improve the accuracy, timeliness, and relevance of financial information. Through the implementation of digital business cases, organisations can achieve improved data accuracy, enhanced decision-making capabilities, increased operational efficiency, and better alignment of financial goals with business objectives.

While there are challenges associated with implementing performance management digital business cases, organisations can overcome these challenges by investing in data integration and governance frameworks, building analytical capabilities, and addressing resistance to change. By leveraging return-based measures such as ROI and ROE, organisations can evaluate the financial impact of their finance transformation initiatives and make informed decisions to drive sustainable growth.

Through case studies, we have seen the transformative power of performance management digital business cases in finance transformation. These cases have demonstrated significant improvements in financial performance, operational efficiency, and decision-making capabilities. To unlock the potential of finance transformation through performance management, organisations should follow these essential steps.

NOTE

1. https://blog.darwinbox.com/top-performance-management-frameworks#:~:text=Google%2C%20for%20example%2C%20adopted%20the,of%20this%20performance%20management%20theory.

CHAPTER 34

Effective Cost and Profitability Management

I N TODAY'S RAPIDLY EVOLVING business landscape, organisations are constantly seeking ways to drive efficiency, improve decision-making, and enhance financial performance. One powerful solution that has emerged is finance transformation. Finance transformation involves the strategic reimagining and restructuring of an organisation's finance function to optimise cost and profitability management. By embracing finance transformation, businesses can unlock new opportunities, streamline processes, and achieve sustainable growth.

Cost and profitability management are vital aspects of any business. By effectively managing costs, organisations can reduce expenses, increase operational efficiency, and boost profitability.

Zhou (2024) argues that without a robust finance transformation strategy in place, cost and profitability management can become challenging and fragmented (Zhou and Wang, 2024). Digital transformation can help in delivering better understanding and improving profitability to help businesses identify areas of growth, optimise pricing strategies, and make informed investment decisions.

Implementing finance transformation initiatives can yield numerous benefits for cost and profitability management. Firstly, it enables organisations to gain a holistic view of their financial data by centralising and standardising processes. This allows for better analysis and decision-making, as well as improved transparency and accountability.

Vasin et al. (2018) contend that finance transformation can automate routine tasks, freeing up valuable resources for more strategic activities. By leveraging advanced analytics and reporting capabilities, businesses can identify cost-saving opportunities and optimise pricing structures to maximise profitability (Vasin et al., 2018).

Finance transformation encompasses several key components that are essential for effective cost and profitability management. These components include:

Process Optimisation: Process optimisation involves streamlining finance processes, eliminating redundancies, and enhancing efficiency. By automating manual tasks,

DOI: 10.1201/9781003514503-34

such as data entry and reconciliation, organisations can allocate resources more effectively and reduce the risk of errors.

Technology Integration: Integrating advanced technologies, such as artificial intelligence and machine learning, can significantly enhance cost and profitability management. These technologies enable businesses to analyse large volumes of data, identify patterns, and generate actionable insights in real-time.

Organisational Alignment: Finance transformation requires strong alignment between finance and other business functions. By fostering collaboration and communication, organisations can ensure that cost and profitability objectives are aligned with overall business goals.

Implementing finance transformation requires careful planning and execution. Here are some key steps to consider:

Assess the Current State: Begin by assessing the current state of your finance function. Identify pain points, inefficiencies, and areas for improvement. This assessment will serve as the foundation for your finance transformation strategy.

Define Objectives: Clearly define your objectives for cost and profitability management. Set measurable goals that align with your organisation's overall strategy. These objectives will guide your finance transformation initiatives and ensure you stay on track.

Develop a Roadmap: Develop a roadmap that outlines the steps and milestones for your finance transformation journey. This roadmap should include timelines, resource requirements, and key performance indicators to measure progress.

Engage Stakeholders: Engage key stakeholders throughout the finance transformation process. Collaborate with finance professionals, IT teams, and business leaders to ensure alignment and secure buy-in for your initiatives.

Execute and Monitor: Execute your finance transformation initiatives according to your roadmap. Continuously monitor progress, track key metrics, and make adjustments as necessary. Regularly communicate updates to stakeholders to maintain transparency and accountability.

Cost management is a critical aspect of finance transformation. Here are some strategies to effectively manage costs:

Identify Cost Drivers: Identify the key drivers of your organisation's costs. Analyse cost structures, processes, and activities to pinpoint areas where cost reductions can be achieved. This analysis will help you prioritise cost-saving initiatives.

Implement Cost Control Measures: Implement cost control measures to reduce unnecessary expenses. This can include optimising procurement processes, renegotiating contracts with suppliers, and implementing cost-saving technologies.

Foster a Culture of Cost Consciousness: Promote a culture of cost consciousness throughout the organisation. Encourage employees at all levels to identify cost-saving opportunities and reward innovative cost-saving ideas.

Profitability management is essential for sustainable growth. Here are some strategies to enhance profitability:

Analyse Pricing Strategies: Analyse your pricing strategies to ensure they align with market demand and maximise profitability. Consider dynamic pricing models, bundling options, and value-based pricing to optimise revenue streams.

Identify Profitability Drivers: Identify the key drivers of profitability within your organisation. This can include product lines, customer segments, or geographical regions. By understanding these drivers, you can allocate resources effectively and focus on high-profit areas.

Monitor and Adjust: Regularly monitor profitability metrics and adjust strategies accordingly. Implement performance dashboards and reporting tools to track profitability in real-time.

This will enable you to make data-driven decisions and proactively address any issues. To illustrate the impact of finance transformation on cost and profitability management, let's explore two case studies:

Dell: Streamlining Financial Processes

Dell, a multinational manufacturing company, implemented finance transformation initiatives to streamline its financial processes. By centralising data and automating routine tasks, they achieved significant cost savings and improved profitability. The finance team now has access to real-time financial data, enabling them to make informed decisions and identify cost-saving opportunities.

Hiya: Optimising Pricing Strategies

Hiya, a technology startup, utilised finance transformation to optimise its pricing strategies. By leveraging advanced analytics and market insights, they were able to identify the most profitable customer segments and adjust pricing accordingly. This resulted in increased revenue and improved overall profitability.

Finance transformation is supported by a range of tools and technologies that facilitate cost and profitability management. Some popular tools include:

Enterprise Resource Planning (ERP) Systems: ERP systems integrate various financial functions, such as accounting, procurement, and inventory management. These systems provide a centralised platform for financial data and enable better cost tracking and control.

Business Intelligence (BI) Tools: BI tools help analyse and visualise financial data, enabling businesses to gain actionable insights. These tools can generate interactive dashboards, reports, and forecasts, empowering finance professionals to make informed decisions.

Robotic Process Automation (RPA): RPA automates repetitive tasks, such as data entry and invoice processing. By eliminating manual errors and improving efficiency, RPA can significantly reduce costs and enhance profitability.

To ensure effective cost and profitability management through finance transformation, consider the following best practices:

1. Regularly assess and optimise your cost structures.

2. Foster a culture of continuous improvement and innovation.

3. Invest in employee training and development to enhance financial skills.

4. Embrace technology and stay updated with the latest tools and trends.

5. Monitor key performance indicators and financial metrics to track progress.

Finance transformation is a powerful strategy for driving business success through effective cost and profitability management. By embracing finance transformation and implementing the key components discussed, organisations can streamline processes, optimise pricing strategies, and achieve sustainable growth. By leveraging the right tools and technologies, businesses can gain a competitive edge and make data-driven decisions. Finance transformation is not a one-time project but an ongoing journey towards continuous improvement and innovation. Start your finance transformation today and unlock the full potential of your organisation's financial performance.

CHAPTER 35

Business Partnering

Iɴ ᴛᴏᴅᴀʏ's ʀᴀᴘɪᴅʟʏ ᴇᴠᴏʟᴠɪɴɢ world, the finance landscape is undergoing a significant transformation. Traditional finance functions are no longer sufficient to meet the demands of the modern business environment. To stay competitive and thrive in this new era, organisations must embrace finance transformation. One key aspect of this transformation is strategic partnership. By forging strategic partnerships, businesses can unlock new opportunities, enhance their capabilities, and ensure long-term success. In this section, we will explore the role of strategic partnership in finance transformation and how it can revolutionise your business.

Strategic partnerships have become a cornerstone of finance transformation due to their ability to drive innovation, efficiency, and growth. In the past, finance departments were largely focused on transactional and routine activities.

Jones et al. (2020) argue that with the advent of technology and the increasing complexity of business operations, finance professionals are now expected to provide strategic insights, drive decision-making, and contribute to the overall success of the organisation. This shift in expectations requires a new approach to finance, one that is collaborative and value driven (Jones and Scapens, 2020).

Strategic partnerships enable businesses to tap into external expertise, resources, and networks, allowing them to augment their existing capabilities and drive transformational change.

There are several benefits of business partnering in finance transformation. Firstly, strategic partnerships can provide access to specialised skills and knowledge that may not be available internally. By collaborating with external partners, businesses can tap into a diverse pool of talent and expertise, enabling them to address complex challenges and drive innovation.

Secondly, strategic partnerships can enhance operational efficiency and streamline processes. By leveraging the strengths of their partners, organisations can optimise their finance functions, automate repetitive tasks, and free up resources for more value-added activities.

DOI: 10.1201/9781003514503-35

Finally, strategic partnerships can foster growth and expansion. By partnering with other organisations, businesses can enter new markets, access new customer segments, and pursue joint ventures that would be difficult to achieve on their own.

Implementing a strategic partnership in finance transformation requires careful planning and execution. Here are the key steps to consider when embarking on this transformative journey:

Define your Strategic Objectives: Before entering into any partnership, it is essential to clearly define your strategic objectives and align them with your overall business goals. Identify the specific areas where you need support or expertise and outline the outcomes you wish to achieve through the partnership.

Identify Potential Partners: Conduct a thorough assessment of potential partners to ensure they align with your strategic objectives, values, and culture. Look for organisations that have complementary skills, expertise, and resources that can enhance your finance function and drive transformation.

Establish Clear Governance and Communication: Establishing clear governance structures and communication channels is crucial for the success of any strategic partnership. Define roles, responsibilities, and decision-making processes upfront to avoid misunderstandings and conflicts down the line. Regular communication and feedback loops are also essential to keep the partnership on track and address any issues or challenges that may arise.

Develop a Mutual value Proposition: A successful strategic partnership should be mutually beneficial for both parties involved. Develop a value proposition that outlines the benefits each partner will derive from the collaboration. This will help build trust, commitment, and a shared sense of purpose, which are essential for long-term success.

Implement a Robust Performance Measurement Framework: To ensure the partnership is delivering the intended outcomes, it is crucial to establish a robust performance measurement framework. Define key performance indicators (KPIs) and set targets that align with your strategic objectives. Regularly monitor and evaluate the partnership's performance against these metrics and make adjustments as necessary.

Continuously Nurture and Evolve the Partnership: Strategic partnerships are not static entities; they require ongoing nurturing and evolution. Regularly review and reassess the partnership's objectives, processes, and outcomes. Seek feedback from all stakeholders and identify areas for improvement or expansion. By continuously investing in the partnership, you can ensure its long-term success and maximise its impact on your finance transformation journey.

While strategic partnerships offer numerous benefits, they also come with their fair share of challenges and risks. It is essential to be aware of these potential pitfalls and take proactive measures to mitigate them. Here are some of the common challenges and risks in finance transformation through strategic partnerships:

Cultural Differences: When entering into a strategic partnership, organisations with different cultures, values, and ways of working may face challenges in aligning their approaches. It is crucial to invest time and effort in understanding and respecting each other's cultures to foster a harmonious and productive partnership.

Lack of Commitment: For a strategic partnership to be successful, all parties involved must be fully committed and actively engaged. Lack of commitment from any partner can hinder progress, create conflicts, and lead to the partnership's failure. It is essential to establish clear expectations and ensure that all partners are dedicated to the partnership's success.

Coordination and Integration Challenges: Integrating different systems, processes, and ways of working can be complex and time-consuming. Misalignment or lack of coordination can result in inefficiencies, errors, and delays. It is crucial to invest in robust project management and change management practices to ensure a smooth and seamless integration.

Dependency and Control Issues: Strategic partnerships involve sharing resources, knowledge, and decision-making authority. This can create dependency and control issues, especially if one partner becomes overly reliant on the other. It is crucial to establish clear boundaries, roles, and responsibilities to maintain a healthy balance of power and foster a mutually beneficial partnership.

Data Security and Confidentiality: Sharing sensitive financial data and information with external partners carries inherent risks. It is essential to establish robust data security measures, confidentiality agreements, and governance frameworks to protect the organisation's sensitive information and ensure compliance with relevant regulations.

By being aware of these challenges and taking proactive measures to address them, businesses can navigate the risks associated with finance transformation through strategic partnership and maximise the benefits they offer.

To truly understand the impact of strategic partnerships on finance transformation, let's explore a few case studies of organisations that have successfully embraced this approach:

Volkswagen Group: Volkswagen Group, a global manufacturing company, was facing challenges in streamlining its finance processes and driving operational efficiency. They entered into a strategic partnership with an external finance service provider that specialised in process optimisation and automation. Through this partnership, Volkswagen Group was able to streamline its finance operations, reduce manual efforts, and improve accuracy and timeliness. The partnership also provided access to cutting-edge technologies and expertise that enabled Volkswagen Group to drive innovation and transform its finance function.

Serve Robotics: Serve Robotics, a fast-growing technology startup, recognised the need for strategic guidance and financial expertise to support its rapid expansion. They formed a strategic partnership with a venture capital firm that had a deep understanding of the technology industry. This partnership not only provided Serve Robotics with the necessary funding but also gave them access to a network of industry experts, market insights, and strategic advice. As a result, Serve Robotics was able to navigate the challenges of scaling their business successfully and achieve sustainable growth.

These case studies highlight the transformative power of strategic partnerships in finance. By leveraging external expertise, resources, and networks, organisations can overcome challenges, drive innovation, and achieve their financial goals.

To ensure the success of business partnering in finance transformation, it is crucial to follow best practices. Here are some key guidelines to consider:

Establish a Shared Vision and Goals: Clearly define the shared vision and goals of the partnership. Ensure that all partners are aligned and committed to achieving these objectives.

Cultivate Open and Transparent Communication: Foster a culture of open and transparent communication. Encourage all partners to share ideas, concerns, and feedback openly.

Regularly communicate updates, progress, and challenges to maintain trust and collaboration.

Build Strong Relationships: Invest time and effort in building strong relationships with your partners. Foster a collaborative and inclusive environment where all partners feel valued and respected.

Leverage Technology and Data: Embrace technology and leverage data to drive informed decision-making and process optimisation. Implement robust systems and tools that enable seamless collaboration and data sharing.

Invest in Training and Development: Provide training and development opportunities for finance professionals to enhance their skills and knowledge in business partnering. This will enable them to effectively contribute to the finance transformation journey.

Regularly Evaluate and Adapt: Continuously evaluate the partnership's performance against agreed-upon metrics. Identify areas for improvement and adapt your approach accordingly. Embrace a culture of continuous learning and improvement.

By following these best practices, businesses can maximise the value of business partnering in finance transformation and ensure long-term success.

Technology plays a crucial role in supporting finance transformation through strategic partnership. Here are some key tools and technologies that can enhance the effectiveness of your partnership:

Cloud-Based Collaboration Platforms: Cloud-based collaboration platforms enable seamless communication, document sharing, and real-time collaboration among partners. They provide a centralised repository for all project-related information and facilitate efficient workflow management.

Data Analytics and Visualisation Tools: Data analytics and visualisation tools enable finance professionals to extract meaningful insights from vast amounts of financial data. These tools can help identify trends, patterns, and anomalies, enabling data-driven decision-making and performance optimisation.

Robotic Process Automation (RPA): RPA automates repetitive and rule-based tasks, freeing up finance professionals to focus on value-added activities. By automating manual processes, RPA improves efficiency, accuracy, and compliance.

Artificial Intelligence (AI) and Machine Learning (ML): AI and ML technologies can be leveraged to automate complex financial processes, enhance forecasting accuracy, and improve risk management. These technologies can also enable intelligent automation and predictive analytics, driving proactive decision-making and strategic insights.

Blockchain Technology: Blockchain technology provides a secure and transparent platform for financial transactions and record-keeping. By leveraging blockchain, organisations can enhance the security, efficiency, and auditability of their financial processes.

By embracing these tools and technologies, organisations can unlock new possibilities, streamline their finance functions, and drive finance transformation through strategic partnership.

To effectively contribute to finance transformation through business partnering, finance professionals need the necessary skills, knowledge, and competencies. Here are some training and development opportunities that can enhance their capabilities:

Business Partnering Workshops and Seminars: Participating in workshops and seminars focused on business partnering can help finance professionals develop a deep understanding of the role and responsibilities of a business partner. These sessions can provide insights into effective communication, relationship-building, and value creation.

Finance Technology Training: Training programmes focused on finance technologies, such as data analytics, RPA, AI, and blockchain, can equip finance professionals with the necessary skills to leverage these technologies in their roles.

Leadership and Influencing Skills Development: Business partnering requires strong leadership and influencing skills. Training programmes that focus on leadership, negotiation, and effective communication can enhance finance professionals' ability to drive change and influence stakeholders.

Continuous Learning and Professional Certifications: Encourage finance professionals to pursue continuous learning and professional certifications in areas relevant to business partnering, such as management accounting, strategic finance, and business analytics. These certifications can provide a solid foundation and demonstrate their commitment to professional growth.

By investing in training and development opportunities, organisations can empower their finance professionals to effectively contribute to finance transformation through business partnering.

As the finance landscape continues to evolve, strategic partnership has emerged as a game-changer for finance transformation. By leveraging external expertise, resources, and networks, businesses can unlock new opportunities, enhance their capabilities, and ensure long-term success. Through strategic partnerships, organisations can drive innovation, streamline processes, foster growth, and navigate the complexities of the modern business environment. However, successful business partnering requires careful planning, clear communication, and a commitment to shared goals. By following best practices, leveraging technology, and investing in training and development, organisations can maximise the benefits of strategic partnerships and embrace the future of finance.

CHAPTER 36

Boosting Employee Performance in the Age of Hybrid Working

I N TODAY'S RAPIDLY EVOLVING business landscape, the way we work has been transformed. The rise of *hybrid working*, combining remote and office-based work, has become the new norm. As organisations adapt to this changing environment, it is crucial to understand how *employee performance* can be optimised in the age of hybrid working. Chief Finance Officer (CFO) must help the organisation provide employees with whatever they need to be productive when remote. One powerful tool that can drive this optimisation is finance transformation. CFOs must consider the benefits, challenges, and strategies of finance transformation in boosting employee performance. They must consider the tools and technologies that can support finance transformation in a hybrid working environment. This can then provide a comprehensive understanding of how finance transformation can enhance employee performance in the age of hybrid working.

Employee performance is a critical factor in the success of any organisation. However, with the shift to hybrid working, new challenges and opportunities have emerged. Understanding the dynamics of employee performance in this environment is essential for organisations to thrive. One aspect to consider is the need for effective communication and collaboration. In a hybrid working model, employees may be physically separated, making it crucial to leverage technology tools that facilitate seamless communication and collaboration. By implementing finance transformation initiatives, organisations can streamline their processes, enable real-time data sharing, and enhance collaboration among teams. This can ultimately lead to improved employee performance.

Another factor to consider is the impact of remote working on employee well-being. While remote work offers flexibility and autonomy, it can also lead to feelings of isolation and burnout. Shwedeh et al. (2023) contend Finance transformation can play a key role in addressing these challenges. By automating mundane tasks and optimising workflows, finance

DOI: 10.1201/9781003514503-36

transformation reduces the burden on employees, allowing them to focus on more strategic and fulfilling work. This can contribute to higher job satisfaction and overall employee performance (Shwedeh et al., 2023).

Finance transformation brings a multitude of benefits that directly impact employee performance. One of the significant advantages is increased efficiency. By automating repetitive and time-consuming tasks, such as data entry and report generation, finance transformation frees up valuable time for employees to focus on high-value activities. This not only improves productivity but also allows employees to utilise their skills and expertise more effectively, leading to a higher level of job satisfaction and performance.

Furthermore, finance transformation enables real-time and data-driven decision-making. With the right tools and technologies in place, organisations can access accurate and up-to-date financial information, empowering employees to make informed decisions promptly. This quick access to critical data eliminates delays and improves the overall agility of the organisation. Employees can respond to market changes swiftly, identify trends, and seize opportunities, thereby enhancing their performance and contributing to the organisation's success.

Finance transformation also fosters a culture of transparency and accountability. By centralising financial processes and implementing robust controls, organisations can ensure that employees have clear visibility into their tasks and responsibilities. This transparency promotes accountability, as employees understand how their contributions align with the organisation's objectives. This sense of purpose and accountability motivates employees to perform at their best, driving overall performance.

While finance transformation offers numerous benefits, it is not without its challenges, especially in the age of hybrid working. One significant challenge is ensuring seamless access to tools and data for remote workers. In a hybrid environment, employees may work from various locations and devices, making it crucial to implement secure and user-friendly technology solutions. Organisations must invest in cloud-based platforms and robust cybersecurity measures to ensure that employees can access the necessary tools and data securely and efficiently. Overcoming these technological challenges is vital to ensuring a smooth finance transformation and maximising employee performance.

Another challenge is maintaining effective communication and collaboration in a hybrid working model. With employees physically scattered, it becomes essential to leverage collaboration tools that facilitate virtual meetings, document sharing, and real-time communication. Organisations must provide training and support to ensure that employees can effectively utilise these tools and collaborate seamlessly. By addressing these communication challenges, organisations can foster a collaborative and engaged workforce, ultimately leading to improved employee performance.

To harness the power of finance transformation in boosting employee performance, organisations should adopt several strategies. First and foremost, it is crucial to prioritise employee well-being. While finance transformation can improve efficiency and productivity, organisations must also ensure that employees do not become overwhelmed or burnt out. By promoting work-life balance, providing mental health support, and fostering a culture of well-being, organisations can enhance employee performance and satisfaction.

220 Finance Transformation

Additionally, organisations should invest in continuous learning and development opportunities. Finance transformation often involves the implementation of new technologies and processes. Providing employees with the necessary training and resources to adapt to these changes is essential. By upskilling employees and fostering a learning culture, organisations can empower employees to embrace finance transformation and leverage its benefits to enhance performance.

Furthermore, organisations should encourage cross-functional collaboration and communication. In a hybrid working model, silos can easily form, hindering collaboration and knowledge sharing. By fostering a culture of collaboration, organisations can break down these barriers and enable employees to work together effectively. This collaboration not only improves employee performance but also drives innovation and problem-solving.

Implementing *finance transformation* in a hybrid working environment requires careful planning and execution. Organisations should start by conducting a thorough assessment of their existing processes and identifying areas for improvement. This assessment should involve all relevant stakeholders, including finance teams, IT departments, and employees. By involving employees in the process, organisations can ensure buy-in and gather valuable insights.

Once the assessment is complete, organisations can develop a roadmap for finance transformation. This roadmap should outline the goals, milestones, and timelines for the transformation process. It should also take into account the unique challenges and requirements of a hybrid working environment. By breaking down the transformation into manageable phases and setting realistic expectations, organisations can ensure a smooth transition and minimise disruptions.

Throughout the implementation process, organisations should provide adequate training and support to employees. This includes training on new tools and technologies, as well as ongoing support to address any challenges or questions that may arise. By investing in comprehensive training and support, organisations can empower employees to embrace the changes brought about by finance transformation and maximise their performance.

Several organisations have successfully implemented finance transformation initiatives in a hybrid working environment, leading to improved employee performance. One such case study is *Wells Fargo*, a multinational operating in the financial services sector. By implementing cloud-based finance systems and automation tools, Wells Fargo streamlined their financial processes, reducing manual effort and improving accuracy. This enabled their finance team to focus on strategic analysis and decision-making, leading to enhanced employee performance and increased value.

Another case study is *Apple*, a technology firm. Apple implemented finance transformation by adopting an integrated financial management system and leveraging data analytics. This allowed them to gain real-time insights into their financial performance and make data-driven decisions promptly. As a result, their employees became more agile and proactive, leading to improved performance and competitiveness in the market.

These case studies highlight the transformative power of finance transformation in a hybrid working environment. By embracing the right tools and technologies, organisations can unlock new levels of employee performance and drive success in a rapidly changing business landscape.

To support finance transformation and enhance employee performance in a hybrid working environment, organisations can leverage a range of *tools and technologies*. One such tool is cloud-based financial management software. Cloud platforms offer the flexibility and scalability necessary for remote work, enabling employees to access financial data and collaborate seamlessly from anywhere. Cloud-based solutions also provide robust security measures, ensuring that sensitive financial information remains protected.

Another technology that can support finance transformation is robotic process automation (RPA). RPA enables organisations to automate repetitive and rule-based tasks, such as data entry and reconciliation. By implementing RPA, organisations can reduce errors, improve efficiency, and free up employees' time for more strategic and value-added activities. This automation not only enhances employee performance but also contributes to cost savings and process optimisation.

Data analytics and business intelligence tools are also instrumental in supporting finance transformation and driving employee performance. These tools enable organisations to analyse vast amounts of financial data, identify trends, and generate actionable insights. By empowering employees with data-driven decision-making capabilities, organisations can enhance their agility and competitiveness in the market.

To fully understand the *impact* of finance transformation on employee performance in a hybrid working environment, organisations should establish relevant metrics and measurement frameworks. These metrics can include both qualitative and quantitative indicators. Qualitative indicators may include employee satisfaction surveys, feedback sessions, and performance reviews. These indicators provide insights into employees' perceptions and experiences regarding finance transformation and its impact on their performance.

Quantitative indicators can include productivity metrics, such as time saved on manual tasks, error rates, and revenue growth. By tracking these metrics over time, organisations can assess the tangible impact of finance transformation on employee performance and overall business outcomes. This data-driven approach allows organisations to make informed decisions and continuously improve their finance transformation initiatives to maximise employee performance.

As organisations navigate the age of hybrid working, finance transformation emerges as a powerful tool for boosting employee performance. By streamlining processes, enabling real-time data sharing, and fostering a culture of collaboration, finance transformation enhances employee productivity, satisfaction, and overall performance. However, implementing finance transformation in a hybrid working environment comes with its own unique challenges. Organisations must prioritise employee well-being, invest in training and support, and leverage the right tools and technologies to ensure a smooth transition and maximise the benefits of finance transformation. By embracing finance transformation, organisations can thrive in the age of hybrid working and drive success in the ever-evolving business landscape.

CHAPTER 37

Hire, Retain, and Develop Digital Skills

I N TODAY'S RAPIDLY EVOLVING digital landscape, the finance industry is undergoing a transformative shift. The integration of technology has revolutionised the way financial institutions operate, making *digital skills* an essential requirement for finance professionals. However, there exists a significant *digital skills gap* within the finance industry, hindering the potential for a seamless transformation.

The digital skills gap refers to the disparity between the digital skills required by organisations and the skills possessed by the current workforce. As the finance industry becomes increasingly reliant on technology, traditional skillsets are no longer sufficient to meet the demands of the digital era. This gap poses a challenge for finance professionals, as they must adapt and acquire new digital skills to remain relevant and contribute to the finance transformation.

To bridge the digital skills gap in finance, it is crucial to understand the importance of digital skills in finance transformation. Finance transformation is the process of leveraging technology and digital capabilities to enhance financial operations, streamline processes, and ultimately drive business growth. Digital skills play a pivotal role in this transformation, enabling finance professionals to navigate complex financial systems, analyse vast amounts of data, and make data-driven decisions.

In today's data-driven economy, the ability to harness the power of technology and digital tools is paramount. Finance professionals equipped with digital skills are better equipped to adapt to changing market dynamics, automate repetitive tasks, and improve efficiency within their organisations. These skills empower finance professionals to move beyond traditional roles and become strategic partners, providing valuable insights and contributing to the overall success of the organisation. By embracing digital skills, finance professionals can unlock a world of opportunities and drive finance transformation within their organisations.

To succeed in finance transformation, finance professionals must acquire and develop specific digital skills. These skills are essential for navigating the digital landscape and

DOI: 10.1201/9781003514503-37

leveraging technology to drive innovation and efficiency. Some key digital skills needed for finance professionals include:

Data Analytics: Data analytics is a critical skill for finance professionals in the digital age. The ability to collect, analyse, and interpret financial data allows finance professionals to gain valuable insights into the financial health of their organisations. By leveraging data analytics tools and techniques, finance professionals can identify trends, detect anomalies, and make informed decisions based on data-driven insights.

Automation and Robotic Process Automation (RPA): Automation and RPA are revolutionising finance operations by streamlining processes, reducing errors, and improving efficiency. Finance professionals with skills in automation and RPA can create automated workflows, develop bots to perform repetitive tasks, and implement systems that optimise financial processes. These skills not only save time and resources but also free up finance professionals to focus on strategic initiatives.

Cybersecurity: As the finance industry becomes increasingly digital, cybersecurity is of paramount importance. Finance professionals must possess a strong understanding of cybersecurity principles and best practices to safeguard sensitive financial data. This includes knowledge of encryption, secure authentication, and data privacy regulations. By integrating cybersecurity into their skillset, finance professionals can protect their organisations from cyber threats and ensure the integrity of financial systems.

Bridging the gap is vital, so acquiring digital skills is a journey that requires dedication, continuous learning, and a proactive approach. To bridge the digital skills gap in finance, finance professionals can consider the following strategies:

Continuous Learning and Professional Development: The digital landscape is constantly evolving, and finance professionals must keep up with the latest trends and technologies. Engaging in continuous learning and professional development is key to acquiring digital skills. This can include attending workshops, webinars, and seminars, enrolling in online courses, or pursuing certifications in relevant digital areas. By investing in their own skill development, finance professionals can stay ahead of the curve and actively contribute to finance transformation.

Collaborative Learning and Knowledge Sharing: Learning from others is a powerful way to acquire digital skills. Finance professionals can leverage internal networks, industry associations, and professional communities to connect with peers and experts in the field.

Collaborative learning environments facilitate the exchange of knowledge, best practices, and innovative ideas. By actively participating in these communities, finance professionals can learn from others' experiences and gain insights into digital skills applicable to finance transformation.

Hands-on Experience and Shadowing: Acquiring digital skills requires practical experience. Finance professionals can seek opportunities to work on digital projects, collaborate with cross-functional teams, or shadow colleagues who possess the desired digital skills. Hands-on experience allows finance professionals to apply theoretical knowledge, gain confidence in utilising digital tools, and develop a deeper understanding of how digital skills can drive finance transformation. By actively seeking out and embracing these opportunities, finance professionals can accelerate their learning and bridge the digital skills gap.

Digital skills are not only integral to finance transformation but also play a crucial role in enhancing finance capabilities. By leveraging digital skills, finance professionals can elevate their contributions and drive meaningful change within their organisations. Some crucial ways digital skills enhance finance capabilities include:

Enhanced Financial Analysis: Digital skills enable finance professionals to analyse financial data more efficiently and accurately. With the ability to leverage data analytics tools and techniques, finance professionals can uncover valuable insights, identify patterns, and make data-driven recommendations. This enhances the accuracy of financial analysis and empowers finance professionals to provide strategic guidance to key stakeholders.

Streamlined Financial Processes: Digital skills allow finance professionals to streamline financial processes through automation and optimisation. By leveraging automation tools and techniques, finance professionals can eliminate manual, repetitive tasks, reduce errors, and improve efficiency. This frees up valuable time and resources, enabling finance professionals to focus on more strategic initiatives and value-added activities.

Improved Collaboration and Communication: Digital skills facilitate seamless collaboration and communication within finance teams and across departments. By leveraging digital tools and platforms, finance professionals can work collaboratively on projects, share information in real-time, and streamline communication channels. This enhances cross-functional collaboration, improves productivity, and accelerates decision-making processes.

Several organisations have successfully undergone finance transformation by leveraging digital skills. These case studies serve as inspiration and provide valuable insights into the power of digital skills in finance transformation.

Henkel: Henkel, the world's largest producer of adhesive technologies, from Persil and Purex in its laundry business unit to Schwarzkopf and Dial in its beauty care vertical, generates over $20 billion annually. It embarked on a finance transformation journey by upskilling its team with digital skills. They invested in training programs and provided hands-on experience focusing on data analytics, automation, and cybersecurity.[1] As a result, they were able to streamline financial processes, enhance data-driven decision-making, and improve cybersecurity measures. This transformation allowed Henkel to stay ahead of the competition, drive business growth, and position itself as an industry leader.

Lloyds Bank: Lloyds Bank, a leading financial institution, recognised the importance of digital skills in finance transformation and implemented a comprehensive digital upskilling program. It provided its finance professionals with opportunities to learn and develop skills in data analytics, automation, and emerging technologies. This upskilling initiative resulted in improved financial analysis, streamlined processes, and enhanced collaboration across departments. Lloyds Bank's finance team became strategic partners, leveraging digital skills to drive innovation, improve customer experience, and achieve operational excellence.

As technology continues to advance, the future of finance transformation is promising and daunting. The finance industry will witness further integration of emerging technologies such as artificial intelligence, machine learning, and blockchain. To navigate this evolving landscape, finance professionals must equip themselves with the necessary digital skills.

The need for digital skills in finance transformation will continue to grow. Finance professionals must be adaptable, open to learning, and proactive in acquiring new digital capabilities. Embracing digital skills is not only essential for personal and professional growth but also crucial for the success of finance transformation initiatives.

Recognising the importance of digital skills, many organisations offer *training and development programmes* to help finance professionals acquire and develop these skills. These programmes can range from in-house workshops to external certifications. Some popular training programmes for acquiring digital skills in finance include:

- Data Analytics for Finance Professionals

- Automation and Robotics in Finance

- Cybersecurity for Financial Institutions

- Digital Transformation in Finance: Strategies and Best Practices

- Emerging Technologies in Finance: A Practical Approach

By enrolling in these programs, finance professionals can gain the necessary digital skills to drive finance transformation within their organisations.

Digital skills are paramount in unlocking the power of finance transformation. The digital skills gap within the finance industry presents a challenge that must be addressed for organisations to thrive in the digital age. By understanding the importance of digital skills, acquiring key digital capabilities, and embracing a proactive approach to continuous learning, finance professionals can bridge the digital skills gap and contribute to successful finance transformation initiatives.

The future of finance transformation holds immense potential, and digital skills will play a pivotal role in shaping this future. By embracing the power of digital skills, finance professionals can drive innovation, improve efficiency, and position themselves as strategic partners within their organisations. It is time to unlock the power of finance transformation through digital skills and embrace the opportunities that lie ahead.

NOTE

1. https://www.accenture.com/bg-en/case-studies/consulting/henkel-cultivates-in-house-talent

Conclusion

The Future of Digital Transformation in Financial Services

DIGITAL TRANSFORMATION HAS BECOME a buzzword in the business world, and the finance industry is no exception. In today's fast-paced digital age, financial services companies are under increasing pressure to adapt and evolve. But what exactly is digital transformation, and why is it so important in the finance industry?

Digital transformation is vital in leveraging digital technologies to fundamentally change the way businesses operate and deliver value to customers. In the context of financial services, it involves using technology to streamline processes, enhance customer experiences, and drive innovation. This transformation is driven by the need to stay competitive in a rapidly changing landscape where customer expectations are higher than ever before.

Digital transformation is no longer a choice for financial services companies; it has become an imperative. The finance industry is facing disruption from fintech startups and tech giants, who are leveraging digital technologies to offer innovative products and services. To stay relevant and competitive, traditional financial institutions must embrace digital transformation.

One of the key benefits of digital transformation in the finance industry is improved operational efficiency. By automating manual processes and digitising paper-based workflows, financial services companies can reduce costs, minimise errors, and increase productivity. This allows them to focus on value-added activities and deliver a superior customer experience.

Digital transformation also opens up new opportunities for revenue growth. By leveraging data analytics and artificial intelligence (AI), financial institutions can gain deep insights into customer behaviour and preferences. This enables them to offer personalised products and services and cross-sell or upsell to existing customers. Additionally, digital transformation enables financial services companies to reach new customer segments, expand into new markets, and create new revenue streams.

While the benefits of digital transformation in the finance industry are clear, there are also several challenges that financial services companies need to overcome. One of the main challenges is legacy systems and infrastructure. Many traditional financial institutions are

DOI: 10.1201/9781003514503-38

built on outdated technology, making it difficult to integrate new digital solutions. Legacy systems are often complex, fragmented, and expensive to maintain, which hinders the pace of digital transformation.

Another challenge is data security and privacy. Financial services companies deal with sensitive customer information, and any breach can have severe consequences. With the increasing frequency and sophistication of cyberattacks, ensuring the security of digital assets and customer data is paramount. Implementing robust cybersecurity measures and complying with regulatory requirements are crucial for successful digital transformation in the finance industry.

Despite these challenges, there are significant opportunities for financial services companies that embrace digital transformation. One such opportunity is the ability to provide seamless and personalised customer experiences. Today's customers expect a seamless omnichannel experience where they can interact with their financial institution through multiple touchpoints, such as mobile apps, websites, and chatbots. By leveraging digital technologies, financial services companies can deliver personalised and relevant experiences that meet customer expectations.

Another opportunity is the ability to leverage data for actionable insights. Financial services companies generate vast amounts of data, from customer transactions to market trends. By using advanced analytics and machine learning (ML), they can extract valuable insights from this data and make data-driven decisions. These insights can be used to identify patterns, detect fraud, assess risks, and optimise business processes. Data-driven decision-making is a key driver of innovation and competitive advantage in the finance industry.

Digital transformation is a complex and multifaceted process that requires careful planning and execution. There are key steps to successful digital transformation in the finance industry. By following these steps, financial services companies can navigate the complexities of digital transformation and unlock its full potential:

Define your Vision and Goals: Before embarking on digital transformation, it is important to have a clear vision and set of goals. What do you want to achieve with digital transformation? Are you looking to improve operational efficiency, enhance customer experiences, or drive innovation? Defining your vision and goals will help guide your digital transformation strategy and ensure alignment across the organisation.

Assess your Current State: Conduct a thorough assessment of your current IT infrastructure, systems, and processes. Identify the gaps and areas for improvement. This will help you prioritise your digital transformation initiatives and allocate resources effectively.

Develop a Roadmap: Based on your vision and goals, develop a roadmap for digital transformation. This roadmap should outline the key initiatives, timelines, and resource requirements. It is important to have a phased approach, starting with quick wins and gradually scaling up.

Build the Right Team: Digital transformation requires a multidisciplinary team with expertise in technology, data analytics, change management, and domain knowledge. Ensure you have the right talent and skills in place to drive your digital transformation initiatives.

Embrace Agile Methodologies: Traditional waterfall project management approaches are no longer effective in the fast-paced digital world. Embrace agile methodologies, such as Scrum or Kanban, to enable iterative and incremental development. This will help you respond quickly to changing market dynamics and customer needs.

Invest in Technology and Infrastructure: To enable digital transformation, you need to invest in the right technology and infrastructure. This may include cloud computing, data analytics tools, customer relationship management systems, and cybersecurity solutions. It is important to choose scalable and flexible solutions that can support your future growth and innovation.

Empower Employees and Foster a Culture of Innovation: Digital transformation is not just about technology; it is also about people and culture. Empower your employees with the necessary skills and knowledge to embrace digital technologies. Encourage a culture of innovation, where employees are encouraged to experiment, fail fast, and learn from their mistakes.

Monitor and Measure Progress: Regularly monitor and measure the progress of your digital transformation initiatives. Use key performance indicators (KPIs) to track the impact of digital transformation on your business metrics, such as customer satisfaction, revenue growth, and operational efficiency. This will help you identify areas of improvement and make necessary course corrections.

Digital transformation in the finance industry is not just about digitising existing processes; it is about leveraging innovative technologies to drive value and create new business models.

Here are some of the key technologies that are driving finance transformation:

AI and ML: AI and machine learning are revolutionising the finance industry. These technologies can automate manual processes, such as credit scoring and fraud detection, and make predictions based on historical data. AI-powered chatbots and virtual assistants are also being used to enhance customer service and provide personalised recommendations.

Blockchain: Blockchain technology has the potential to transform the way financial transactions are conducted. It provides a secure and transparent way to record and verify transactions, eliminating the need for intermediaries. Blockchain can streamline processes, reduce costs, and enhance trust and transparency in the finance industry.

Robotic Process Automation (RPA): RPA involves using software robots to automate repetitive and rule-based tasks. In the finance industry, RPA can be used to automate back-office processes such as data entry, account reconciliation, and regulatory compliance. This frees up employees to focus on more strategic and value-added activities.

Cloud Computing: Cloud computing enables financial services companies to access and store data and applications over the internet rather than on local servers. This provides scalability, flexibility, and cost savings. Cloud computing also enables collaboration and data sharing across different departments and locations.

Internet of Things (IoT): The IoT refers to the network of physical devices, vehicles, and other objects embedded with sensors and software, which enables them to collect and exchange data. In the finance industry, the IoT can be used for asset tracking, risk management, and personalised insurance pricing. For example, insurance companies can use IoT devices to monitor driving behaviour and offer personalised car insurance premiums.

Big Data Analytics: Financial services companies generate vast amounts of data, from customer transactions to market trends. Big data analytics involves using advanced analytics techniques to extract insights from this data and make data-driven decisions. This can help identify patterns, detect fraud, assess risks, and optimise business processes.

By implementing these innovative technologies, financial services companies can drive finance transformation and gain a competitive edge.

In the digital age, customer experience has become a key differentiator for financial services companies. Customers expect seamless and personalised experiences across all touchpoints, from opening an account to resolving a query. Here's how customer experience plays a crucial role in finance transformation:

Enhanced Customer Satisfaction: By leveraging digital technologies, financial services companies can deliver personalised and relevant experiences that meet customer expectations. This leads to higher customer satisfaction and loyalty.

Improved Customer Retention: Providing a superior customer experience can help financial services companies retain their existing customers. Customers are more likely to stay with a financial institution that understands their needs and delivers a seamless experience.

Increased Customer Acquisition: Word-of-mouth recommendations and positive online reviews are powerful drivers of customer acquisition. When customers have a positive experience with a financial institution, they are more likely to recommend it to their friends and family.

Brand Differentiation: In a crowded and competitive market, customer experience can be a key differentiator. A financial institution that offers an exceptional customer experience is more likely to stand out and attract new customers.

Data-driven Personalisation: Digital transformation enables financial services companies to leverage data analytics and AI to offer personalised products and services. By understanding customer preferences and behaviour, financial institutions can tailor their offerings to individual needs.

Omnichannel Experience: In today's digital world, customers expect a seamless omnichannel experience. They want to be able to interact with their financial institution through multiple touchpoints, such as mobile apps, websites, and social media. By providing an omnichannel experience, financial services companies can meet customer expectations and enhance engagement.

Embarking on digital transformation can be a daunting task, but there are several resources and tools available to help financial services companies navigate this journey. Here are some key resources and tools for finance transformation:

Industry Reports and Whitepapers: Stay updated with the latest industry trends and insights by reading reports and whitepapers from reputable sources. These resources provide valuable information on best practices, case studies, and emerging technologies.

Professional Networks and Communities: Join professional networks and communities to connect with peers, share experiences, and learn from each other. These networks provide a platform for collaboration and knowledge sharing.

Digital Transformation Consultants: Consider engaging the services of digital transformation consultants who specialise in the finance industry. These consultants can provide guidance, expertise, and support throughout the digital transformation journey.

Training and Certification Programmes: Invest in training and certification programmes to upskill your employees and equip them with the necessary knowledge and skills for digital transformation. There are several online and offline programmes available that cater specifically to the finance industry.

Digital Transformation Platforms and Tools: There are several platforms and tools available that can help streamline and automate digital transformation initiatives. These tools provide features such as project management, collaboration, and analytics, enabling financial services companies to drive digital transformation more efficiently.

Digital transformation is a critical imperative for financial services companies. By understanding the importance of digital transformation, addressing key challenges and opportunities, and implementing innovative technologies, financial services companies can unlock the full potential of finance transformation. The role of customer experience cannot be underestimated, as it plays a crucial role in driving customer satisfaction, loyalty, and acquisition. Looking ahead, the future of digital transformation in financial services is promising, with trends such as AI, open banking, RegTech, and blockchain shaping the industry.

The finance industry is on an exciting journey of digital transformation, and the future looks promising. Here are some key trends that will shape the future of digital transformation in financial services:

AI and ML: AI and ML will continue to play a significant role in finance transformation. These technologies will enable more advanced analytics, personalised customer experiences, and automated processes.

Open Banking: Open banking is a regulatory initiative that allows customers to share their financial data with third-party providers. This enables customers to access a wider range of financial products and services, and promotes competition and innovation in the finance industry.

Regulatory Technology (RegTech): RegTech involves using technology to streamline regulatory compliance processes. As regulatory requirements become more stringent, financial services companies will increasingly rely on RegTech solutions to ensure compliance and mitigate risks.

Cybersecurity and Data Privacy: With the increasing frequency and sophistication of cyberattacks, cybersecurity and data privacy will be top priorities for financial services companies. They will need to invest in robust cybersecurity measures and ensure compliance with data protection regulations.

Blockchain and Digital Currencies: Blockchain technology and digital currencies, such as cryptocurrencies, have the potential to disrupt traditional banking and payment systems. Financial services companies will need to explore the opportunities and challenges associated with these technologies.

Collaboration with Fintech Startups: Fintech startups are driving innovation in the finance industry, and traditional financial institutions can benefit from collaboration. By partnering with fintech startups, financial services companies can access innovative technologies and reach new customer segments.

The future of digital transformation in financial services is exciting and full of opportunities. Financial services that embrace digital transformation and stay ahead of the curve will thrive in the digital age.

References

Ahmad, Israr. (2024). The Role of Strategic Financial Management in Enhancing Corporate Value and Competitiveness in the Digital Economy. Economía Chilena. 1–08. 10.36923/economa. v27i1.116.

An, Shijie. (2023). The Impact of Digital Transformation on the Financial Ser-vices Industry: A Comprehensive Review. Advances in Economics, Management and Political Sciences. 30. 36–41. 10.54254/2754-1169/30/20231417.

Avira, Silvia & Rofi'ah, & Setyaningsih, Endang & Utami, Suryandari. (2023). Digital Transformation in Financial Management: Harnessing Technology for Business Success. Influence: International Journal of Science Review. 5. 336–345. 10.54783/influencejournal.v5i2.161.

Aznag, Fatma & Tahanout, Keira. (2022). IoT Solutions for Fintech and Banking Industry. 16. 745–763.

Charkha, Sanket. (2023). Financial Specialists' Needs, Challenges, as well as Potential opportunities Amid the Transformation to Digital Financing.

Fu, Tao & Li, Jiangjun. (2023). An Empirical Analysis of the Impact of ESG on Financial Performance: The Moderating Role of Digital Transformation. Frontiers in Environmental Science. 11. 10.3389/fenvs.2023.1256052.

Gao, Tianyue. (2024). Fintech: Digital Transformation in the Financial Industry. Advances in Economics, Management and Political Sciences. 74. 1–7. 10.54254/2754-1169/74/20241863.

Girardone, Claudia & Kokas, Sotirios & Wood, Geoffrey. (2021). Diversity and women in finance: Challenges and future perspectives. Journal of Corporate Finance. 71. 101906. 10.1016/j. jcorpfin.2021.101906.

Gunturu, Srinivasa Rao & Godbole, Madhavi & Josyula, Hari Prasad. (2024). FinTech - Automatic Payment Process in the ERP System. International Journal of Computer Trends and Technology. 72. 99–103. 10.14445/22312803/IJCTT-V72I11P116.

Hakizimana, Samuel & Wairimu, Muraguri & Muathe, Stephen. (2023). Digital Banking Transformation and Performance-Where Do We Stand?. International Journal of Management Research and Emerging Sciences. 13. 10.56536/ijmres.v13i1.404.

Harsono, Iwan & Suprapti, Ida. (2024). The Role of Fintech in Transforming Traditional Financial Services. Accounting Studies and Tax Journal (Count). 1. 81–91. 10.62207/gfzvtd24.

Huang, Meng & Gao, Sen. (2024). Digital Transformation Strategy for Financial Management of Entity Enterprises in the Information Age. Applied Mathematics and Nonlinear Sciences. 9. 10.2478/amns-2024-0533.

Ilsøe, Anna & Karma, Kadri & Larsen, Trine & Larsson, Bengt & Lehr, Alex & Masso, Jaan & Pavlenkova, Ilona & Rolandsson, Bertil. (2022). The Digital Transformation of Financial Services Markets and Industrial Relations. https://faos.ku.dk/publikationer/forskningsnotater/rapporter-2019/Rapport_187_-_The_digital_transformation_of_financial_services_markets_and_industrial_relations.pdf

Jones, J., & Scapens, B. (2020). Finance Business Partnering: Design Principles to Orchestrate Value. (4 ed.) (CIMA Research Executive Summary). https://www.cimaglobal.com/Research-Insight/Finance-business- partnering-Design-principles-to-orchestrate-value/

Kamuangu, Paulin & K K, Paul. (2024). Digital Transformation in Finance: A Review of Current Research and Future Directions in FinTech. World Journal of Advanced Research and Reviews. 21. 1667–1675. 10.30574/wjarr.2024.21.3.0904.

Kapotas, Spyridon. (2023). Digital Transformation of Organizations and Their Organizational Cultures: A Case Study in a National Defense Industry. Saudi Journal of Economics and Finance. 7. 442–458. 10.36348/sjef.2023.v07i10.004.

Khashabi, Pooyan & Kretschmer, Tobias. (2019). Digital Transformation and Organization Design – A Complex Relationship. SSRN Electronic Journal. 10.2139/ssrn.3437334.

Lekhi, Pooja. (2023). Web 3.0 Revolution in the Finance Industry: Exploring Blockchain and Decentralized Finance. 10.4018/978-1-6684-9919-1.ch002.

Mousa, Saeed & Bouraoui, Taoufik. (2023). The Role of Sustainability and Innovation in Financial Services Business Transformation. Theoretical Economics Letters. 13. 84–108. 10.4236/tel.2023.131005.

Olaoye, Godwin & Daniel, Samon. (2024). Role of Enterprise Resource Planning (ERP) in Digital Transformation. https://www.researchgate.net/publication/378299651_Role_of_Enterprise_Resource_Planning_ERP_in_Digital_Transformation

Papathomas, Aristides & Konteos, George. (2023). Financial Institutions Digital Transformation: The Stages of the Journey and Business Metrics to Follow. Journal of Financial Services Marketing. 10.1057/s41264-023-00223-x.

Pelykh, Vitaliy. (2020). Finance 4.0 as the Idea of Digital Transformation of the Financial Sector. World of Economics and Management. 20. 134–148. 10.25205/2542-0429-2020-20-2-134-148.

Qi, Yanli. (2023). Fintech and the Digital Transformation of Financial Services. Highlights in Business, Economics and Management. 8. 322–327. 10.54097/hbem.v8i.7225.

Robert, Abill. (2024). The Role of Diversity and Inclusion in Talent Acquisition Strategies. International Journal of Management.

Saxunova, Darina & le Roux, Corlise. (2020). Digital Transformation of World Finance. 10.5772/intechopen.93987.

Shwedeh, Fanar & Aburayya, Ahmad & Mansour, Muntaser. (2023). The Impact of Organizational Digital Transformation on Employee Performance: A Study in the UAE. Migration Letters. 20. 1260–1274.

Siddiqui, Ayesha & Yadav, Arti & Farhan, Najib. (2023). Digital Transformation of Financial Services in the Era of Fintech. 10.1002/9781119905028.ch2.

Su, Xin & Wang, Shengwen & Li, Feifei. (2023). The Impact of Digital Transformation on ESG Performance Based on the Mediating Effect of Dynamic Capabilities. Sustainability. 15. 13506. 10.3390/su151813506.

Thottoli, Mohammed & Islam, Md. Aminul & Yusof, Mohd & Hassan, Md. Sharif & Hassan, Md Arif. (2023). Embracing Digital Transformation in Financial Services: From Past to Future. SAGE Open. 13. 10.1177/21582440231214590.

Vasin, Sergey & Gamidullaeva, Leyla & Vardan, Mkrttchian. (2018). Transformation of the Cost Management System for Implementing Business Projects in Digital Entrepreneurship. SHS Web of Conferences. 55. 01012. 10.1051/shsconf/20185501012.

Viswanathan, Balachandran. (2022). Digital Transformation in Finance -Need, Challenges and Opportunities for Professionals. https://www.researchgate.net/publication/358573483_Digital_Transformation_in_Finance_-Need_Challenges_and_Opportunities_for_Professionals

Vunjak, Nenad & Zelenović, Vera & Birovljev, Jelena & Milenković, Ivan. (2012). Strategic Planning in Banking. Technics Technologies Education Management. 7. 196–203.

Wang, Le. (2023). Fintech: Digital Transformation in Finance. Advances in Economics, Management and Political Sciences. 40. 22–27. 10.54254/2754-1169/40/20231983.

Wang, Xiao. (2024). Research on the Digital Transformation of the Financial Industry. Highlights in Business, Economics and Management. 24. 1786–1791. 10.54097/582g8r55.

Zhou, Shouliang & Wang, Zixuan. (2024). Transformation of Financial Accounting to Management Accounting in the Era of Artificial Intelligence. Frontiers in Computing and Intelligent Systems. 7. 33–37. 10.54097/wph0vg88.

Index

Milton Keynes UK
Ingram Content Group UK Ltd.
UKHW011829161024
449632UK00008B/274